MW00527779

The Anthropocene Cookbook

The Anthropocene Cookbook

Recipes and Opportunities for Future Catastrophes

Zane Cerpina and Stahl Stenslie

The MIT Press

Cambridge, Massachusetts

London, England

The Anthropocene Cookbook is supported by TEKS—Trondheim Electronic Arts Centre

TRONDHEIM ELECTRONIC ARTS CENTRE
WWW.TEKS.NO

The MIT Press would like to thank the anonymous peer reviewers who provided comments on drafts of this book. The generous work of academic experts is essential for establishing the authority and quality of our publications. We acknowledge with gratitude the contributions of these otherwise uncredited readers.

This book was designed and set in Adobe Caslon Pro by Zane Cerpina. Printed and bound in the United States of America.

Library of Congress Cataloging-in-Publication Data

Names: Cerpina, Zane, author. | Stenslie, Stahl, author.
Title: The anthropocene cookbook : recipes and opportunities for future catastrophes / Zane Cerpina and Stahl Stenslie.
Description: Cambridge, Massachusetts : The MIT Press, 2022. | Includes bibliographical references and index.
Identifiers: LCCN 2022001049 (print) | LCCN 2022001050 (ebook) | ISBN 9780262047401 (hardcover) | ISBN 9780262371636 (ebook) | ISBN 9780262371643 (pdf)
Subjects: LCSH: Cooking. | Sustainable living. | Food habits—Forecasting. | Food—Environmental aspects. | Human beings—Effect of environment on. | Nature—Effect of human beings on. | Geology, Stratigraphic—Anthropocene.
Classification: LCC TX714 .C457 2022 (print) | LCC TX714 (ebook) | DDC 641.5—dc23/eng/20220611
LC record available at https://lccn.loc.gov/2022001049
LC ebook record available at https://lccn.loc.gov/2022001050

10 9 8 7 6 5 4 3 2 1

This book is dedicated to the interesting times we live in. This book and the many wondrous projects discussed herein could be made only because of the ecological crises of the Anthropocene.

Book Disclaimer

This disclaimer is for all the food ideas, recipes, and ingredients discussed in *The Anthropocene Cookbook*.

The authors disclaim liability for incidental or consequential damages and assume no responsibility or liability for any loss or damage suffered by any person as a result of the use of the information in the book. The authors assume or undertake no liability for any loss or damage suffered as a result of the use of any information found in the book.

This book is intended solely for discussions about and reflections on the future of food. It is not intended for step-by-step cooking. If you decide to experiment with the ingredients mentioned in this book, you are doing so at your own risk. We make no warranties for the outcome of your own food experiments. The authors of this book are not liable or responsible for any loss or injury from your interpretation or use of the information discussed in this book.

This book does not provide any medical or dietary advice. The authors of this book are not liable for any adverse reactions (including food poisoning, food allergies, or food-borne disease) or any other negative outcome resulting from the use of the recipes and information in this book.

Contents

Introduction

Introduction

You cannot prohibit the catastrophe, you must surf it!
—Paul Virilio[1]

A large flood of probable catastrophes looms over us in the age of the Anthropocene—the Age of Man. In this new geological epoch, it is we humans who are reshaping the planet, impacting geology and ecosystems alike.[2] The number of self-inflicted doomsday scenarios grows by the day, confronting us with an endless chain of world-ending armageddons. No wonder the end of humanity appears nigh. Or? This coming era can also be seen as an opportunity for change, new ideas, and a constructive future. Focusing on foods we soon might eat, *The Anthropocene Cookbook* is all about seeing and seizing these sizzling moments that lie ahead of us. For certain, the Anthropocene will put our thinking, creativity, and ingenuity to the test. Let us welcome that and throw ourselves heads first into the waves of change.

Recipes for Future Catastrophes

While this cookbook presents several tasty and nutritious, albeit alternative, menus, the book's subtitle is not without a double meaning. On the one hand, with a future gone sideways and known food sources uncertain, we do need new recipes for thinking about, producing, and consuming food. On the other, it should not come as a surprise that the current climate crisis is partly caused by the cookbook itself. There is a boundless demand for the most exquisite foods, promoted by the more than ninety thousand traditional cookbooks for sale in bookstores and online:

The typical early twenty-first-century cookbook, with its gorgeous illustrations, elegiac combinations of the failing fruits of the Earth with those that cost us the climate, water, soil, and our safety to produce is, unambiguously, a recipe for disaster on a planetary scale.
—Julian Cribb[3]

Seen from this perspective, any cookbook indulges us into resource depletion, willingly and happily eating our earth empty and way beyond the planetary boundaries. Thus, every cookbook hastens the end of the world that it is written in. So too with *The Anthropocene Cookbook*. Whether it helps delay the coming crisis or accelerates it, cooking for the future is no easy task.

Cooking for the Future

The Anthropocene Cookbook investigates humanity's future cuisine in a time when humankind shapes, models, and engineers the earth beyond recognition. The book poses the two following questions about our menu to be: How can we survive in an age of constant environmental catastrophes? And how then to thrive?

The answers feature the most innovative ideas and speculative concepts about potential future foods conceived within the fields of art, design, science, technology, philosophy, and food studies. *The Anthropocene Cookbook* rethinks our eating habits and traditions, challenges food taboos, and proposes new recipes for our survival in times of dark ecological disasters. Here more than sixty projects give insights and thought-provoking ideas while offering a radical new take on our future cuisine.

This book is about eating for our future survival but also about exploring all the mouthwatering experiences on the way. The end of food as we know it does not mean the end of eating or the end of a good and tasty diet. In the current and lasting crisis, we need a roadmap to our edible futures. To get there, we urge visionary thinking and original suggestions. After all, it is our future that we are dealing with. In the Anthropocene, this is a pressing issue as our extinction as a species is a possible scenario.

A Different Cookbook

Cookbooks are most often written in a pretty straightforward manner. They start by introducing a more or less specialized gastronomic topic. Then they list the ingredients and guide the reader line by line through composing, mixing, handling, cooking, and even serving the dishes of one's choosing. Like Marinetti's *Futurist Cookbook*, it takes a good cookbook to cook up more than just food.[4] And it takes much more than a collection of recipes when it comes to preparing for future catastrophes.

This is no conventional cookbook. There are no ordinary recipes to be found herein either. We cannot expect our future to materialize based on step-by-step instructions. For this reason, *The Anthropocene Cookbook* is more a tool book for creative actions than a collection of traditional recipes. It is a necessary celebration of the inventive, out-of-the-box thinking that we will need in the Anthropocene. The content here, accordingly, explores how to think and act differently through the art of cooking.

Food concerns us all every day, and it is, therefore, a raw material well suited for art and cultural commentaries. This cookbook is composed of a number of extraordinary proposals and concepts relevant for thinking about our futures through food. More than edibility is on this menu. Radical and powerful imaginations are necessary to develop solutions for surviving the large numbers of looming catastrophes expected to hit humanity within a few generations.

On Methodology
The overarching strategy in this cookbook is to advance thinking in and about the Anthropocene. To do so, the book endeavors to delve into a wide range of artistic concepts, methods, and works as its main approach. It is a cookbook of ideas exemplified through the presented projects. The aim is to build communicable future narratives through these artistic and creative practices, both material and conceptual. In that way, it can be associated with ventures such as UNESCO's Futures Literacy project, which wants to build our skills to better understand the role that the future plays in what we see and do.[5] One of its major differences from other future investigative projects is that the research presented in this cookbook is, first and foremost, artistic in its nature and methodology. The majority of the selected projects are from the visual arts—a field rife with the transgressions and queer thinking necessary to foster an unimaginable imagination. Many of the cutting-edge projects in this book are also rooted in the fields of design, science, industry, science fiction, and film.

Although the metamethodology is to research the Anthropocene through these selected works, the individual methodologies of these projects vary. The works are wildly diverse not only in their creative approaches but also often in their agendas. What they do share are the layered and multifaceted characteristics common to many groundbreaking practical and aesthetical experiments. They are often hybrid, cross-pollinated, and constructed through bricolage and interdisciplinary methods.

The projects presented here are not chosen because of their nutritional value or their delight to the palate or sensual significance alone. They are selected because they represent new and previously unimagined takes on how to make and depict food, both for consumption and ideation. They propose radical ideas through the practice of cooking. If they do not cook what they preach, they somehow use food as a medium for critical and—more often than not—subversive and thought-provoking thinking. If there is one quality we need to develop for the many unforeseeable crises in the Anthropocene, it is our ability to imagine the inconceivable, the unlikely, even the unwanted. That is the most valued trait in a transitional period when we are about to step into unknown territories. And liminal transition needs liminal thinking.

Liminal Aesthetics

As we stand on the threshold of a new Age of Man, we are in need of new ethics and new aesthetics that are fit for this transitory and liminal time. The proposals and projects in this book both border on a new world and urge our thinking through art. They are therefore objects of liminal aesthetics: the aesthetics produced and experienced in the transitory phase as we move through ambiguous and uncertain times. While this book is unlikely to fill your stomach through tasty foods, many of the concepts presented are productive in the sense they function as building blocks for new ideas and values.[6] They question our beliefs, influence the ways we think, stimulate creative inquiries, and widen our territory of action. In that sense, they are also transgressive, taking us out of our comfort zone into unknown domains and potential discoveries. So they serve well as roadmaps to a bumpy future we do not know but whose contours already challenge our capacity to both imagine and create.

With a future gone dark, it is all the more imperative to make the impossible possible. To do so, artistic thinking is a perfect partner in crime as it is individualistic and idiosyncratic in all its flavors of alternative practices. Frequently, these appear to go against or even contradict our habitual understanding of the world. Given the open and disparate nature of art, this undertaking is necessarily an open-ended, often paradoxical, dissonant, and conflicting approach to the question of how to survive in the age of catastrophes. The advantage of seeing the Anthropocene through the lens of art is the frequent combination of observation with actions in surprising and—as with the presented works—often unforeseen ways. Given the slippery task of guessing concrete futures, art is the field of choice to prepare us for the unpredictable.

Liminal aesthetics is a necessity for precarious times, representing an ambiguous, disoriented, even confused transitory phase when standing on the verge of shifting from one understanding to another. Thus, liminal aesthetics also represents a rite of passage. The liminal methodology in this book is deduced from our focus on artworks relevant to and within this transitory phase in time. A further outcome of a liminal situatedness is the consequential transgressive transition. Transgressive actions in art can function as a phenomenological breakdown of our instrumental understanding of the world.[7] Such breakdown scenarios promote reflection. With knowledge shattered, reflection on the catastrophe sinks in. Thus, one can see the aim of liminal or extreme art proposals as the attempt to produce necessary reorientations.

While art and artistic methodologies are the main ingredients, this book uses methodologies and frameworks from several fields. Many of the presented recipes for imagination and new aesthetics are inspired by critical and discursive design thinking that emphasizes the ethical and societal features of design. Critical design is here understood as the shift away from the production of commercial objects toward an interrogative, experimental, and discursive practice.[8] Discursive design also has a main focus on the discourse produced by objects and, in the case of this book, can be "good(s) for thinking."[9] Such design objects can be purely speculative yet highly successful in commenting futures, both desirable and undesirable.[10] So they provide a ground for reflective thinking rather than concrete solutions. That resonates with a future that is a moving target at best. If liminal aesthetics prepares us for the unexpected, we also need to prepare for the extreme.

Black Swans in Extreme Scenarios

The Anthropocene will hit humanity hard and in ways we cannot always foresee. What we do know is that these unforeseeable events will be of a *black swan* nature—unpredictable, improbable situations that impact the course of our lives.[11] On the compelling side, black swan events always challenge us on the spot to think outside the box and come up with viable solutions to sudden existential obstacles.

This itself is a good reason to investigate our futures through art. Artistic thinking is put at the center of this book as—to quote Giorgio de Chirico—the object of art "is to create previously unknown sensations; to strip art of everything routine and accepted."[12] Deep familiarity and affinity with the unknown might be the best preparation for strange and uncertain futures.

How do we deal with a future without food as we know it? A failure to resolve such a future crisis could be nothing less than a failure of imagination. This further underlines the role of transgression—as the sudden transition from a normative good into something extremely different, even unknown—as one of the methods in this book.

The coming age of catastrophes presents us with acute challenges. What role can art play in these times? Can art inspire innovative solutions while under the pressure of an extreme scenario?[13] Or do the arts—as many might expect—produce only eccentric, unrealizable concepts? Providing operable solutions to extreme scenarios is not necessarily the aim. Within the plethora of artistic ideas and imaginations, there is a hope that we will stumble upon not just the instrumental, functional, or interesting but also the transgressive that opens up new spaces of possibilities for living and thinking.

The examples in this book can be seen as extreme food scenarios whose primary goal is not to satisfy your hunger or your taste but to initiate thinking about a future gone wild. Playing out and testing extreme scenarios can break down habitual thinking patterns and understanding of the world today. The consequence of facing unknown futures is that in time, sooner or later, its unfolding will lead to the breakdown of previous understandings. Those are the black swans in extreme scenarios whose contours can be traced in the coming chapters.

Book Chapters

The Anthropocene Cookbook is organized around different food types, cooking methods, and taste preferences. The first chapter, Cooking for Survival, is about how humans have developed the historically complex relationships we have to our food and environment. Food is what communities and societies are built on, and lack of it has made many fall apart. The past has many lessons for humanity as it tries to learn to survive in the Anthropocene. Civilizations such as the Mayans, Easter Islanders, and the Greenland Norse vanished after first draining all their natural resources and then failing to adapt to the conditions they had created for themselves. There are also the dangerous progress traps—new advances that appear beneficial at first sight and yet have long-term negative consequences. We have continued to fall for progress traps ever since our ancestors first perfected hunting tools and soon drove whole species into extinction.

Introduced are also some immense threats to future global food security. Ongoing events—mass extinction of species, deforestation, desertification, and water scarcity—are about to render traditional food production impossible. Many culturally rooted problems must be acknowledged. New eating disorders, obsessions with ecological consumerism, and always abundant food at the hunting ground (also known as a supermarket) are some of the reasons that the future food crisis appears inconspicuous. Further, the chapter points toward necessary adaptations to the changing environmental conditions as a measure to avoid future famines and even the collapse of civilization. Putting survival back at the top of the agenda underlines the need for new ways to produce and consume food.

Chapter 2, Ecological Crisis Menu, offers a new take on crisis staple foods and alternative approaches to food security during times of ecological disasters. The diet of the past is marked by the human ability to source alternative foods in times of emergencies and extreme scenarios. This chapter takes knowledge of past famine meals as a useful ingredient when developing the future diet.

One of the challenges explored in chapter 2 is how to transform the traditional approach of stockpiling to ensure a food supply for future crises. For example, artist Tattfoo Tan in his project *New Earth Meals Ready to Eat* (*NEMRE*) (2012) proposes turning the food waste of today into future emergency provisions. However, even if we continuously stockpiled the entire global food surplus, it would be only a short-term solution. Extreme disasters such as the interruption of sunlight due to nuclear or volcanic winter would affect food production for decades. Even without the occurrence of such extreme scenarios, human-driven environmental changes are enough to severely disrupt global food security for equally long periods. The time-tested famine staples will not be enough to ensure the survival of the global population in the Anthropocene.

The chapter further envisions our soon-to-be menu through alternative agriculture. Food production could even take place in the cities, where almost

70 percent of the global population will be living by the year 2050. Food utopias such as rooftop gardens of lettuce and strawberries will not feed the world sufficiently. However, closed-loop aeroponic and aquaponic farms could allow for the production of many nutritious staples. And even if this came at a high energy cost, indoor agriculture might be the only viable solution under unstable environmental conditions in the Anthropocene.

Chapter 2 demonstrates that we will need to develop new foods and invent entirely new ways of eating. We could consume untapped existing edible resources such as plankton and earthworms. An even more radical suggestion is to utilize foods that are considered inedible today. This could be achieved by expanding human digestive abilities through the use of synthetic biology. Artist Gints Gabrans explores such a scenario in the project *FOOOD* (2014), which proposes giving humans the ability to synthesize an enzyme that can break down cellulose. If humans could gain the full nutritional value from grass, wood, and even paper, nutritious meals would always be at hand. With a similar method, rotten food, including rotten meat, could appear to taste even more delicious than fresh food.

Alternative scientific ways to produce nutrition with a minimal amount of resources are also presented. If trees were to vanish and all the rotten food we had stockpiled eaten, we could use bacteria to make food out of air. Another speculative idea is to achieve partial human photosynthesis. Despite the enormous environmental crises threatening food security, the artistic contributions presented in chapter 2 show that we can create and adapt to new nutritious foods and previously unknown flavors. The ecological crisis has arrived, and it does not necessarily mean we will have less to eat—just different.

Chapter 3, Anthropocene Specials, explores the new, potentially edible and delicious ingredients that have emerged due to human-induced changes in the environment in the new Age of Man. What new and innovative opportunities have we created for ourselves? Artworks presented in the chapter offer diverse perspectives on the food in the Anthropocene. Some projects focus on exposing the human role in shaping the earth and its habitat. Others dive directly into exploring the culinary possibilities and potential advantages of new raw ingredients and resources.

Until recently, our diet has been characterized only by foods of natural origin. Now, many animals and plants are being genetically engineered to increase the yield and otherwise improve their value as food for humans. As demonstrated in chapter 3, one case stands out in particular—the domesticated chicken. The breeding and modifying of the broiler chicken to meet our insatiable appetite for meat is likely to turn the massive deposit of chicken bones into a geological marker in the earth's crust.

More often, we encounter stories with quite the opposite outcome. The unfolding Anthropocene extinction is likely to take some of our favorite dishes off the menu in the near future. The project *GhostFood* (2013) by Miriam Simun and Miriam Songster explores how to preserve the tastes that will be lost to extinction. The project simulates the experience of eating soon-to-be-extinct foods by pairing synthetic scents and facsimile edible ingredients.

The Anthropocene also brings many new resources to add to the menu, including invasive species. Eating invasives is promoted to eliminate the non-native organisms that threaten the local biodiversity, food production, and sometimes human health. However, as chapter 3 shows, in many cases, the arrival of invasive species can boost local economies and enrich diets.

Besides new organic resources, there are also new human-made raw materials. Plastic, radiation, and various types of pollution enter and gradually become part of our environment and food systems. While initially ending up in our stomachs by accident, in the future, these ingredients might eventually be welcomed in our diet. Until now, they have remained unfathomably large hyperobjects. However, several artworks and design projects described in the chapter anticipate their potential role in our food system, traditions, and daily rituals. Although invisible, human-made radiation is already part of our diet both unintentionally and intentionally through the use of radiation breeding to improve crops. Plastics can become a nutritious feed for organisms (such as meal moths, fungi, and bacteria) that can be further turned into nutritious edibles for humans. And air pollution can be used to spice up foods and capture the true taste of the Anthropocene.

Also, water pollution contains potentially valuable resources. As Jon Cohrs and Morgan Levy propose in the project *Alviso's Medicinal All-Salt* (2010), polluted wastewaters can be used to reharvest various useful compounds. The project results in a hand-harvested salt enriched with pharmaceuticals, including antibiotics and antidepressants from local wastewaters.

The chapter demonstrates how the new raw ingredients of the Anthropocene become visible, tangible, and even edible when exposed and investigated through art and design. Investigations into the unlikely, unfamiliar, often preposterous resources present rewarding ways forward toward adaptation and survival in the Anthropocene. There are also ample reasons to rethink the difference between foods and nonfoods.

Chapter 4, Fake Foodies, explores our abilities to mimic traditional dishes by replacing the primary food ingredients with alternative and often synthetic sources. If consuming traditional foods of organic origin will no longer be viable or culturally accepted in the Anthropocene, they might be replaced with fake foods. In the transition toward the artificial and fake, these novel foods are being introduced to be as close as possible to the original dish. Thoroughly different resources can be used to mimic some traditional foods perfectly.

Urban legends about fake foods from around the world make it a controversial topic. Imposter foods have traditionally been feared by consumers. Some faux foods, like cheese produced from umbrella handles and eggs made from synthetic resin, have turned out to be hoaxes. Others, such as synthetically watered milk produced in China, have presented serious health hazards resulting in fatalities.

On the other side, there is evidence of potential consumer acceptance of faux foods. One example is the Chinese wine counterfeiting industry, confirming how the taste for fake can be culturally acquired. Fake foods can potentially become more familiar than the original and even preferred. Imposter foods have already gained acceptance as replacements for animal-based meat. Why eat meat if we can work toward meat substitutes that taste better, are healthier, and eliminate the negative environmental costs associated with the meat industry? The project *Sea-Meat Seaweed* (2016) by Hanan Alkouh proposes a meatless future based on seaweed while problematizing the transition away from meat-keeping procedures and rituals. In a post-meat world, the roles of farmer, slaughterer, and butcher might be lost. This loss of cultures could negatively impact the fabric of our society, norms, and values.

The chapter covers the vast amount of artistic, design, and scientific contributions that explore the most advanced technologies to produce imposter meat through cellular agriculture. The prospect of lab-grown meat comes with various products grown from cells of farm animals, frogs, fish, and even insects. Art plays an important role when uncovering the technological and ethical issues of this emerging biotechnology and its potential impact on the food industry and the environment. *Tiger Penis Project* (2018) by artist Kuang-Yi Ku investigates whether a mix of cellular agriculture, 3D printing, and gene modification could replace the illegal market of animal parts used in traditional Chinese medicine. With a recent milestone of the first approved lab-grown meat to be sold commercially in Singapore, even lab-grown human meat might become a commercial possibility.

Chapter 5, Human Deli, continues exploring the culinary possibilities offered by the human body. The projects presented in the chapter explore how the body can be utilized as a resource for new and exquisite locally sourced flavors instead of exotic spices from across the world. The human body is full of edible ingredients varying in flavors, textures, aromas, cooking qualities, and even nutritional values. Most of the artistic contributions discussed in chapter 5 are feasible hands-on kitchen and lab experiments with human bodily materials.

Organic human materials such as semen, breast milk, and bacteria can spice up our diet and even contribute to important processes in food production. Biologist Christina Agapakis and olfactory researcher Sissel Tolaas in the project *Selfmade* (2013) explore the role of human bacteria in the cheese-making process. *Selfmade* is a collection of cheeses made from human bacterial starter cultures using samples from human hands, feet, noses, and armpits. While the cheeses are not intended to be edible, the authors speculate that the presence of human bacteria has resulted in unique cheese flavors and aromas. On a similar note, several creative projects studied in chapter 5 have used probiotic bacteria from the human vagina to create yeast starter cultures used in bread, yogurt, and beer production. A similar hands-on approach has been taken to explore the potential use of human breast milk. Such experiments have led to the creation of breast milk cheese and even commercially sold ice cream. The use of human breast milk in food production also uncovers ethical dilemmas different from those that arise from the farming of animals. While human breast milk can be obtained with consent, it is intended to feed the newborn.

Other bodily matters, such as human feces, can play a more significant role as a nutritional food supplement, substantially contributing to future food security. In human feces, there is not only precious water but also undigested material, including bacterial biomass and proteins that can be re-eaten. Even human urine can be recycled into clean drinking water and other valuable resources such as sugar and unabsorbed medicine.

Recent methods can turn most human excess materials into potential resources. The enormous amount of human material removed through plastic surgeries might be turned into useful proteins. We could also grow human meat in the lab, giving an innovative spin to cannibalism. This method would provide victimless protein without the risk of transmittable diseases that occur through other cannibalistic practices. Artist Theresa Schubert in her project *mEat me* (2020) explores such a scenario by growing muscle cells taken from a biopsy of her own thigh. The artist is concerned with the environmental impact of industrial farming and simultaneously explores and critiques the clean-meat promises of the in vitro meat industry.

If future diets are to be nutritious and yet full of new exotic culinary experiences, human by-products can be valuable. However, humans are not only eating. They are also being eaten, which makes for an interesting question about how to make us taste better for the nonhuman species that consume us. By contributing as food for others, we can feed such organisms as insects that, in turn, might become an important food source for humans in the future.

Chapter 6, Bug Buffet, takes a critical look at the potential for insects to replace animal protein sources. Insects might be the ultimate protein source due to their high nutritional value, fast growth, and low environmental impact—at least in theory. There is also the potential to reduce organic waste by using it as feed for farmed bugs. And even better, mealworms that can digest Styrofoam could turn nonorganic waste into nutritious food for humans. Not less important is the wide variety of flavors and textures that the nearly one million different insect species living on the earth can offer. They present tastes we are already familiar with, including lemongrass, pork, fish, nuts, and potatoes.

With the long list of advantages offered by insects, it might seem that the only thing standing in the way of turning this sustainable resource into a global industry is social acceptability, especially in Western countries. Design, packaging, and branding can play essential roles in bringing this traditional food back to tables around the world. The designerly projects showcased in chapter 6 offer specialized tools for harvesting, farming, and eating insects. Innovative technologies can also contribute to making the food itself visually more desirable. *The Insects Au Gratin* (2011) project by Susana Soares reinvents insect cuisine using insect paste to 3D print unique food objects. The final product is a series of aesthetical edible sculptures that have no resemblance to bugs.

However, even if the taboo of eating bugs is overcome and all required legislation is set in place, insects are not compatible with everyone's dietary needs. Potential for food allergies and chemical contamination can repel consumers from eating bugs. Breeding and eating insects might also be objected to by the vegan community and others who see eating insects as unethical as eating products from farmed animals.

The chapter also investigates the feasibility of providing enough insect-based foods for the global population. A significant drawback is the seasonal and limited availability of insects in temperate and northern climate zones. And like many other natural resources, the global insect populations face a worrying decline in the Anthropocene. This could be solved by farming insects in controlled environments both on an industrial scale and directly in urban households. Among the many projects posing a solution to insect farming is the *Hive* (2018) by the LIVIN Farms studio. The *Hive* is a kitchenware product that allows its users to become microlivestock farmers.

There are several potential technological obstacles to farming insects in a sustainable way. The heating, cooling, air conditioning, and air humidity needed for rearing insects in temperate climates consume large amounts of energy. If bugs are ever to become a primary protein source for the world, the experiment needs to be undertaken on a global scale and will have uncertain outcomes. Given experiences with upscaling animal farming, there is no guarantee that microlivestock production will succeed on a planetary level.

With the Western countries struggling to appreciate insects as food, the experiment of farming bugs instead of beef is likely to be postponed. For now, insects seem to remain on the menu as alternative emergency food. While the world struggles to swallow bugs, the age-old race to discover other perfect food sources continues.

Chapter 7, Future Superfoods, explores the unceasing dream of creating foods with enormous or even magical powers. Ancient Greeks told legends about ambrosia—the gods' food that would provide immortality. Science fiction brought us the idea of a meal in a pill—a small tablet that provides complete nutritional fulfillment. And food healthism has led to heightened interest in superfoods and an endless flow of health benefit claims for almost any food. It is unlikely that curry will save your life or that broccoli can undo diabetes. Even worse, the focus on food healthism can lead to orthorexia—an eating disorder originating from an unhealthy obsession with eating healthy food.

Future superfoods will have to fulfill many obligations. Not only should they contribute to the survival and health benefits of the human species, but they also should enhance our eating experiences and set a new course for the relationship between humans and nature. While the meal in a pill remains a techno-utopian dream, new and often intriguing proposals continue to emerge. Many of them, including protein-rich cockroach milk, are unlikely to ever be produced in large quantities.

The chapter further examines the trend of upgrading existing foods into superfoods. In the case of golden rice, genetic engineering of the rice crop promises to end vitamin A deficiency. It is likely that many superfoods in the future will be simply an enhanced version of staples we know today. Even farm animals that have already been heavily redesigned can be further upgraded to increase their yield and change characteristics of meat and dairy. Artist Paul Gong in his project *The Cow of Tomorrow* (2015) speculates that farm animals could even be used to produce electricity. Gong proposes implementing a tiny turbine in the artery of the animal, allowing it to use blood flow to harness energy.

Or we can enhance our own bodies with superfoods instead. The *Algaculture* (2010–) project by Michael Burton and Michiko Nitta proposes merging human organs with algae to enable a symbiotic, semiphotosynthetic relationship with one of the most nutritious foods on earth. However, algae are not fully compatible with the human body due to allergies and other health risks they pose, even if just consumed as a food.

While it might not be one single food, the search for the ultimate superfood opens doors to many new and exciting journeys into taste. Emerging technologies also open up the possibility of creating personalized superfoods. The Japanese design studio Open Meals has created a restaurant concept called *Sushi Singularity* (2019) to explore hyperpersonalization in the food industry. *Sushi Singularity* will serve personalized food based on the guests' individual DNA, health assessment, and nutritional needs.

With the endless list of superfoods, it might seem like menus of the future will consist of only perfected superfoods. However, the chapter also acknowledges the human desire to indulge in luscious snacks.

Chapter 8, Future Junk Food, looks at how fast food and the junk-food industry might be changed by adverse environmental conditions in the Anthropocene. Fatty oversized hamburgers, salty french fries, and addictive sugary soft drinks might be the junk foods of our age, but what will be the favorite snacks, healthy or not, for the next generations? How will the ecological crisis and increasing health concerns shape junk-food production in the future?

Whatever it is made of, future junk food will be just as high on our menu as it is today. Modern junk-food items are perfected to the bliss point, the ideal combo of the three irresistible food qualities—sweet, salty, and fatty. This makes people want endlessly more of them. And the consequences are dire. The global population is getting more obese, while the reckless production of junk foods is severely damaging both the environment and natural resources. Although there is an increasing trend toward healthier diets, a global and collective shift toward eradicating junk foods entirely is unlikely. Humans are not genetically programmed to drool at low-calorie green vegetables the same way they do over fatty hamburgers and sugary beverages. For now, and until human nature is reengineered to eat differently, junk-food obsession is here to stay. When envisioning the future diet, we must also incorporate the unhealthy side of it.

With the Anthropocene unfolding, the junk-food industry is bound to be affected by coming environmental crises. To satisfy the growing consumption of junk foods with natural resources depleted, we must consider the use of alternative ingredients. When anything can be made to taste like anything else, as is explored in chapter 4, a circular and sustainable approach would be to consider using resources that are available in abundance. Junk food could even be made out of real junk—from food waste to paper and even plastics. Henry Hargreaves and Caitlin Levin envision an extreme version of such food items in the project *Deep-Fried Gadgets* (2013). The project proposes an electronics-infused cuisine through a series of deep-fried earplugs, laptops, smartphones, and other electronic devices.

In the future, alternative production methods for meat-based junk food and fast food will be in particular focus. Meat substitutes can range from plants, algae, insects, and food waste to lab-grown meat.

Modern junk foods have succeeded in satisfying the natural cravings of humans. Nonetheless, junk food is much more than just ingredients. It is also the food fantasies that many of these foods are built around. In an age marked by endless environmental crises, including the unfolding sixth mass extinction of species, it can be challenging to let our imaginations wander freely and wildly. Are fantastic food futures even possible in the age of constant ecological crises?

Chapter 9, Fantastic Cuisine, presents the most bizarre and extravagant meals that might someday become part of our diet. Food fantasies are an essential tool in shaping our edible futures. A product of our imagination today might become a real meal tomorrow. Some yet untasted organic flavors might be brought from other planets. However, it is unlikely that the most fantastic food items in the future will be provided directly by Mother Nature. Our soon-to-be cuisine does not have to be bound to natural resources. New technologies can be useful to source and develop entirely new mythical foods.

The chapter looks into how genetic engineering, cellular agriculture, and de-extinction open up unexplored possibilities for food production. In the future, we might bring back legendary foods from the past. Our menu could consist of dinosaurs, mammoths, passenger pigeons, and other species that our ancestors ate to extinction. The Center for Genomic Gastronomy in *The De-extinction Deli* (2013) questions the risks and outcomes of de-extinction becoming a reality. Would we enter a paradoxical loop and bring back extinct species just to eat them into extinction over and over again? This will also come with new ethical dilemmas about which species, if not all, to resurrect. If so, will they come back as invasive species impacting current habitats beyond the new normal?

Moving away from the obsession with foods from the past, we might also extract DNA from existing species for recombination into new and mythical ones such as lab-grown centaurs and mermaid meat. The myths and fantasies of eating unicorns and dragons could become an everyday thing in the future. The project *Eating E.T.—Mock Alien BBQ* (2014) by studio Unsworn and artist Terje Östling offers edible fantasies of alien organisms from outer space. The barbeque of a life-size, gluten replica of E.T.—the famous science fiction character from the movie *E.T. the Extra-Terrestrial*—questions why we choose to eat some species but consider it unethical to eat others.

Fast-forwarding humanity's constant chase for better and tastier meals, the ecological catastrophes are bound to taste delicious. Whatever the future might hold, our imagination is what pushes the food industry and innovation ahead. Failure to solve a crisis is a failure of imaginations.

Celebrating Imagination

A cookbook is much more than just a collection of recipes. It is a staple item in almost every kitchen around the world. A cookbook symbolizes the traditions and values of society. After all, "Tell me what you eat, and I shall tell you what you are."[14] This book signalizes new traditions for the new you. It also introduces new foods that will set future cuisine trends that will be valued for generations. *The Anthropocene Cookbook* imagines the meals to come. It is a cookbook that takes you for a wild and delicious ride into our unknown futures.

Cooking for Survival

1 Cooking for Survival

The struggle for food has always been the chief preoccupation of Mankind.
—John Boyd-Orr[1]

The history of food is the history of humanity. More than anything, food is at the top of the human agenda. Our survival has always depended on our ability to find, select, and store food. Hunting and gathering were daily obligations for our ancestors. Our omnivore nature, allowing us to consume food of both plant and animal origins, has been an essential factor in our persistence and our existence as the species we are today. Our adaptive diet not only ensured our survival but also changed our bodies biologically over time and drastically shaped the environment around us.

Between 2 million and 3.9 million years ago, the diet of early humans shifted dramatically.[2] The use of new, more nutritious, and energy-dense plants channeled necessary calories into accelerating the development of the human brain.[3] Later the increasing meat consumption is thought to have ensured further expansion and transformation of the brain[4]—the vital organ that in turn powered our abilities to innovate better tools for hunting and gathering food.

Our evolution goes hand in hand with our diet and food choices. The cultural-historical hypothesis of milk drinking is that some lucky farmers evolved the right genome to digest the white gold as the cow was domesticated some eight thousand years ago. This trait soon expanded to larger human populations.[5] Humans evolved to keep the lactase enzymes active into adulthood, making dairy a staple food.[6] Cooking was another crucial component that contributed to our evolution biologically and, in the words of Levi Strauss, helped "mark the transition from nature to culture."[7]

The new food choices brought about in the Anthropocene are likely to ush-er in a whole new era for our kitchens and transform our bodies even further. History shows that "humans are always on the lookout for something better, bigger, tastier."[8] The Anthropocene might just be the next big and unexpected leap in our cuisine that we have been craving for.

Historical Collapses of Societies

Food has had a massive significance in all aspects of human history. Food is what communities and societies are built on, and lack of it makes them fall apart. The collapses of past societies often follow a similar pattern: population growth led to intensified agricultural production and farming, eventually resulting in unsustainable practices that increasingly damaged the environment. Severe food shortages, starvation, and fighting over these declining resources are the last straws breaking apart any community in crisis. This follows the Malthu-sian model of population collapse: when an exponential population growth crosses the (often) linear production growth of food, then disintegration and catastrophe follow.[9]

Current projections of a rapidly growing global population combined with environmental threats to our linear growth of a food supply suggest we are heading in the same direction as "the civilizations of the Maya, the Anasazi, the Pitcairn and Easter Islanders, and the Greenland Norse."[10] These great civilizations vanished after draining all their environmental resources and then failing to adapt to the conditions they had created for themselves.

The Anthropocene Epoch

The Anthropocene designates the new geologic epoch that we are currently in. It is defined by the human species' massive geological impact on the planet. The name is composed of the Greek words *anthropo* (man) and *cene* (new). Coined by Eugene Stoermer and Paul Crutzen in 2000, it describes the effect of humanity as a major geological force. The meanings for *Anthro-pocene* differ across scholarly communities. There is also debate as to when the era started, but the advent of the nuclear bomb is a definitive and mea-surable marker in the earth's crust. The Anthropocene Working Group therefore agreed that the Anthropocene epoch began in the early 1950s with the first thermonuclear tests and consequently deposits of artificial radionuclides on earth.

Our impact on the earth is now so large that a faraway future investigation into the earth's crust would uncover not just radioactivity but a clearly defined layer of sedimentary rock (strata) left behind from all human activity. Should an end to the Anthropocene stratum be uncovered, then that discovery would pose an existential question as to why the epoch stopped. It might well signify the abrupt end of human civilization, possibly even of our species. That expectancy is one of the reasons that the Anthropocene has become synonymous with the age of catastrophes. The changes to our home world are happening at a more rapid rate than ever before in previous epochs. Our carbon-fueled activity, in particular the release of carbon dioxide (CO_2) by the combustion of fossil fuels, is currently shaping both the climate and ecosystems of the earth. The effect of this could be described as a wild ride toward Hothouse Earth. The global effects are no longer invisible hyperobjects but are increasingly felt locally and individually through warmer temperatures, melting ice, rising sea levels, and progressively violent weather. We are unlikely to slow down our planet-shaping activities in the next decades. Therefore, their effects are expected to brutally escalate. A dire prognosis of havoc-wreaking life conditions on earth is just one result. That is why the Anthropocene has become synonymous also with the age of extinction. The balance of earth is about to become so upset that life will not be able to adapt quickly enough to the changing conditions. Human-caused heating of the oceans and deforestation of the lands have effects similar to a scorched-earth policy that destroys anything that is good for life as we know it, resulting in a catastrophic loss of species. Less biodiversity and a destitute biosphere will impact food production and resources negatively, thus impacting our future menus in ways we cannot or prefer not to foresee. We therefore have to think catastrophically about our future cuisine.

Global Food Concerns

Anthropocene cuisine will be strongly defined by changing environmental conditions. Our access to resources will always impact what we can put on the table. Current threats to our global food security are immense. The sixth mass extinction is threatening earth's wildlife, including plant and animal species we use in our food supply.[11] Deforestation, desertification, and water scarcity limit the availability of potentially usable and already in use farmland for food production. Meanwhile, overfishing and aquatic pollution are pushing the limits of ocean resources.

Some of the causes that will put the quality and quantity of food produced globally under dire pressure are less obvious than others, especially those with a far-reaching domino effect. One such example is the decline of insects that will result in a decrease in birds, bats, reptiles, amphibians, fish, and even small mammals.[12]

Having fewer insects also threatens the harvest of fruits and vegetables. Approximately a third of the world's food supply consists of crops that need to be pollinated by insects.[13] While many important insect species are dying out, the increasing abundance of pests is stripping the fields of essential food crops completely barren. Invasive species such as the Pacific oyster and the Asian giant hornet bring unpredictable changes to biodiversity by taking over and replacing the native ones (see chapter 2, Ecological Crisis Menu, and chapter 6, Bug Buffet). Adding to this comes our appetite for consuming the rare and nearly extinct. Animals in the rear-view mirror may appear as more plentiful than they are. Likewise, a great success today can lead to a disaster tomorrow.

The Progress Trap

Our food habits and eating patterns impact what we produce to secure the global food supply. Innovations and new discoveries also impact changes in the availability and quality of the edible resources on earth. Their implementation is often perceived as progress toward better eating and guaranteed survival. That is not always the case, at least not from a long-term perspective.

How are we to know if advances in the production of food that we consider beneficial for our diet today are also for the better in future phases of the Anthropocene? What happens if our perceived successes have negative consequences? Such an instance can be described as a progress trap: "a chain of successes which, upon reaching a certain scale, leads to disaster."[14] Throughout history, humans have fallen into many progress traps because of food. The first progress trap in history is thought to have occurred when humans perfected hunting methods that in turn drove whole species to extinction: "Palaeolithic hunters who learnt how to kill two mammoths instead of one had made progress. Those who learnt how to kill 200—by driving a whole herd over a cliff—had made too much. They lived high for a while, then starved."[15]

We have never learned from the mistake of the first progress trap. Humans have since hunted and harvested many animal and plant species to extinction. Steller's sea cow was eaten up and away by hungry Russian hunters in the

1700s.[16] Bluefin tuna and sharks are some of the species eaten to the brink of extinction today, the latter being the main ingredient for a traditional Chinese soup made through the illegal harvest of shark fins.[17]

Why do we keep falling into the same progress trap? Although on a moral level, we often advocate sustainability as the protection of endangered species, our everyday struggles often uncover paradoxical yet realistic views on how we actually go about it. This is illustrated by the following story: when a fisherman realizes the local fish sources are overexploited and running out, he does not stop. Instead, he just speeds up his fishing for extinction. Asked about his son's future, he leaves a simple message: "he'll have to find something else."[18]

A progress trap we gladly embrace today is the salmon farming industry. Although it provides the world a lot of important fish oils and proteins, it comes at a considerable cost. Salmon are carnivorous creatures that feed on smaller fish. The production of each kilogram of farmed salmon therefore demands at least two kilograms of other fish. Those meals come from the catching of wild and "nonfood" fish such as types of anchovies and sardines.[19] The rapid growth of the salmon farms has led to a fishing race that swiftly depletes the earth's oceans and results in a net loss of marine resources. The science and success behind farming salmon to satisfy the world's hunger for the tasty fish comes at a very high price, surrendering environmental sustainability to a needy palate and cash profit. Our impression of progress is trapped in a cannibalistic loop of reasoning with an end in sight.

Early humans escaped the first progress trap of overhunting by inventing farming.[20] It ended the Old Stone Age and has been seen as an improvement that not only ensured survival but also advanced culture. However, an escape from one progress trap can be a temporary success leading to an even greater pitfall.[21]

Modern agriculture might be a delicate trap we have set for ourselves. Our current diet is strongly defined by the same few crops that ancient people learned to cultivate in the late Stone Age.[22] While it can be seen as a sign of success in selective breeding, the focus on farming only a handful of staple crops in large quantities without rotation can result in a disaster. Monocultures are fragile. Their lack of genetic variability means that a single new pest or disease can wipe out the entire crop population. This was the case with the Irish potato famine between 1845 and 1850, when a potato blight triggered starvation that killed or displaced 25 percent of the Irish population.[23] If our current farming methods head us into a new progress trap, the question is, then, What will the Anthropocene generations have to eat except for the leftovers of today?

Coming Famines

History repeats itself. Famines (extreme shortages of food leading to mass starvation) have tormented all societies on some level in the past. Between the 1860s and 2016, it is estimated that 128 million lives were lost to starvation.[24] Compared to earlier historical periods, fewer people have died in famines in recent decades.[25] Today, lack of food and malnutrition are significant threats only in the poorest countries or locally affected areas around the world, often as a result of human error. Similar to contemporary famines, the most infamous cases of mass starvation have been products of political or military pressure, such as the Great Leap Forward famine in China from 1958 to 1962 and the more recent famine during the Syrian civil war.[26] Yet, despite this common belief, the vast majority of famines throughout history have been caused by changing environmental factors.[27]

The complex environmental crises of the Anthropocene might be more than enough to lead the world into a new era of global starvation. Our contemporary food production practices and eating habits are further fueling the future food crisis.

The Hunting Ground Supermarket

Our relationship with food is now very different from that of early humans. Hunting and gathering have never been easier. For most people living in developed countries, everything from the global garden of earth is available at the reach of a hand. The supermarket has now become our new forest.[28] With the goods harvested around the world, we fill our stomachs with anything we desire around the clock. It appears as if we have an endless food supply to continue feeding the growing world population.

Suffering the burden of surplus and a full stomach, it can be difficult to imagine the modern world struck by famine. However, the sudden onset of the Covid-19 pandemic has shown that even the developed nations can suffer from the smallest disruption in the global food supply chain.[29]

The existing threats can also be easily overlooked since the world is increasingly affected by severe obesity and the harmful consumption of salt, fats, and sugars. The eating habits in developed countries are characterized by the dangerous Western diet, a modern dietary pattern consisting mainly of high intakes of red meat, highly processed food, salt, and sugar. This diet is increasingly becoming popular around the world, including China, whose 1.4 billion people over the past two decades have developed a taste for fatty and high-protein foods. Their adoption of new eating habits has put an additional strain on the global food supply.[30]

The rich selection of foods in supermarkets worldwide reflect the immense scale of global trade, influencing not only what we can buy but also how we compose and pioneer our diets.

Emerging Eating Disorders

The supermarket brought a disconnect to how food is sourced and gathered and drastically impacts our eating patterns. Instead of a daily search for food, we keep ourselves busy with ever new and particular ways of perfecting our meals. New eats can be peculiar. From consuming only raw foods to practicing juice diets or choosing only single-colored foods, the comfortable and well-fed modern human has developed many strange eating habits that likely would not have crossed the minds of our ancestors. Early humans would have probably preferred junk food and gaining easy energy through oversugared soda and binge eating at McDonald's. But what about more modern luxury eating disorders such as bulimia, anorexia, and even something as fascinating as orthorexia nervosa—an eating disorder caused by an unhealthy obsession with eating healthy food?[31] In line with the symptoms of conventional eating disorders, orthorexia nervosa can manifest itself through "obsessive thinking, compulsive behavior, self-punishment, escalating restriction."[32] Emerging eating behaviors, including orthorexia, are rooted in social and cultural trends as well as the course set by the food industry.

Greenwashing Ecology

The global food choices and food ethics are heavily steered by consumerism. Even the inexpensive and convenient fast-food industry that has dominated the food market is increasingly being pressured to recognize global environmental concerns. This has resulted in a shift toward sustainable food production and a healthy diet as a marketing trick.

The food sector is now progressively focusing on green consumption, which allows it to sell products that can be marketed as green, natural, ecofriendly, and sustainable. The customer is invited to make the right choice through the act of simply consuming. Buying a product can now assure customers that they are caring for the environment and contributing to a better and greener future for all. This sounds like hypocritical and cynical greenwashing, which it often is.[33]

Slavoj Žižek describes the Starbucks syndrome as cleaning your consciousness through buying products that are marketed as good for the environment. By purchasing a product, you can even donate to a cause or charity. Can we really save rainforests and pandas and feed starving kids in Africa merely by drinking an overpriced caffè latte? We do right in questioning the existing eco- and sustainability trends. According to Žižek, "Most of what we think of as radical or subversive—or even simply ethical—doesn't actually change anything."[34] Žižek includes "sustainable living" and "eating eco-friendly foods" in the category of things that make us feel good but bring about no real change.[35]

The ecocriticism philosopher Timothy Morton is further skeptical of what we believe to be sustainable. According to Morton, figuring out how to survive and live in the Anthropocene requires us to radically rethink the concepts of nature and ecology. The main issue here is that the idea of sustainability is built on a romanticized view of nature as something we need to protect and restore to its earlier pristine form. Morton believes that to imagine a truly sustainable future for humanity, we must realize that "Ecology may be without nature. But it is not without us."[36] Dividing the world into natural and unnatural sets up fake borders. It is no longer possible to distinguish between them in this Age of Man. This makes any marketing claims such as "wholesome, all-natural, and farm-fresh" to be a romantic illusion.[37]

Morton further states that when thinking about ecology, we must also include human-made hyperobjects—entities of vast temporal and spatial dimensions, such as plastic bags, global warming, radiation, and pollution.[38] To sustain the soon-to-be 10 billion global population,[39] we can no longer count only on Mother Earth's natural resources. How would our diet look from a new ecological perspective with hyperobjects in the food chain equation?

Food Utopias and Dystopias
The central role of food in the survival of the human species makes it an appealing and exciting topic to speculate on. From science fiction to art, design, research, and public debates, there are plenty of both utopian and dystopian future food visions.

We have dreamt of a food paradise since ancient times.[40] The most optimistic utopias are usually simply marked by an abundance of desired foods. In the opposite direction is the idea of feeding oneself with the fewest resources possible while enjoying all the tastes and gaining the nutritional benefits one would acquire from traditional food. Hence, popular food utopias are often connected to promises of scientific and technological solutions that give high expectations of food production, cooking, and consumption.[41] Any wish to return to Mother Nature and live in a purely natural environment while providing ecological food to everyone should be recognized as one of the most popular modern food utopias.

The current scale of the global environmental crises makes it much easier to imagine a Malthusian and dystopian rather than utopian future on this planet. A wide range of possible extreme events could cause significant and abrupt declines in global food production through conventional agriculture, even leading to the collapse of human civilization. The bleakest future dystopias—such as the 1973 movie *Soylent Green*—portray a world struck by famines where humans are fighting for their survival, consuming strange, last-resort foods made from recycled human corpses.[42]

Uncertain Menus

The thought of hunger makes the imagination work. Long-term adaptation to the Anthropocene requires us to develop a diverse and genuinely sustainable diet using a variety of available resources. The chosen food sources must also be well equipped to withstand hostile climates, environmental conditions, diseases, and other external stresses. Even previously unimaginable dishes could become staples of tomorrow.

Like our ancestors, we will have to be on the move, testing and tasting foods we have never tried before. Consuming an unknown substance can be hazardous, even deadly. Trying new food for the first time can therefore be scary. It evokes neophobia—the fear of the unknown. "No matter what we may gain, our first reactions to new are suspicion, skepticism, and fear. This is the right response: most ideas are bad."[43]

Sometimes even the best food ideas have to fight neophobia. The acceptance of the potato in Europe took a very long time from its introduction in the late sixteenth century. It is a rags-to-riches story where even after decades it was considered a food for the poor at best.[44] It took more than a century for it to be hailed as a crop that not only fed Europe but also helped to avoid the Malthusian trap.

Uncovering new edible and before unknown foods will push our survival further. "Creating something new may kill us; creating nothing new certainly will. This makes us creatures of contradiction: we need and fear change."[45] Trying new foods can be risky, but we are also neophiles: we desire the unknown. A certain dose of wildness and risk-taking will, therefore, come in handy in the future.

Summary

When it comes to survival, food has always been and will continue to be at the top of our agenda. The food choices of our past have had a direct role in the evolution of our brain, physiology, and digestive abilities. In turn, our eating patterns immensely affect the environment and availability of foods to eat next. This makes for a vicious relationship with food. Historically, the majority of famines have been caused by changing environmental factors. Not comprehending the bigger picture of environmental conditions has led to catastrophic food crises and even the collapse of entire societies.

Now, in the Anthropocene, our food concerns are more immense than ever. Not only are food crises coming our way from all directions, but they are also on a global scale. From a past and a present that have seen the extinction of species, water scarcity, and soil depletion, the future brings the challenge to ensure enough food for the survival of the global population. Unlike our ancestors, we have the mechanisms and knowledge to foresee and handle many of the food crises thrusting our way. While collapsed societies of the past did not succeed in adapting their diets to the conditions of environmental disasters, we might be able to. However, our relationship with food is more complicated than ever. The coming famines are disguised by the current abundance and an easy availability of food, global obesity, and strangely luxurious eating disorders. The romanticized view of nature also stands in the way of adapting food production to withstand the coming food emergencies.

To ensure our future survival, we must see beyond the comfortable foodscapes of today. Instead of falling for ecoconsumerism and dreaming of feeding the world only with organic foods provided by Mother Nature, we must develop a new menu fit for the environmental crises of our times.

Ecological Crisis Menu

2 Ecological Crisis Menu

What do you eat when the sky has darkened, temperatures are subzero, winter continues into summer, and crops are gone? When nature, as you know it, stops, dies and mass hunger ensues, how do you survive for the next three years? Such was the scenario during the Fimbulwinter (great winter) that ravaged northern Europe from around 536 to 538. It is estimated that as much as half the population died during this time.[1] The great damage imprinted a lasting memory on its culture. In Norse mythology, the Fimbulwinter refers to the prelude to Ragnarok (the end of the world). One likely cause of its origin was the prolonged solar darkness experienced around the world in 536. There is mounting evidence that this was an aftereffect of a massive eruption of the Ilopango volcano in El Salvador.[2] A next Fimbulwinter is just one of the many potentially large-scale natural and ecological catastrophes that are expected to happen again in our near future—that is, within one thousand years. So how are we to survive in times of deep environmental crisis? How can we do better than just endure and actually eat good, plentiful, and healthy? How are we to eat sufficiently or even in abundance when traditional food sources no longer exist?

In the best near-future scenario, even if the climate on earth stabilizes, we will have to develop and expand our diet to ensure global food security for the growing global population. To achieve this, our next cuisine will need to be seasoned with technological and scientific advances. We have to discover and innovate brand-new edible resources as well as continue to advance and improve already existing foods. After that, when the consequences of the large-scale ecological crises that are crashing our way begin to set in, our diet will again be challenged and require adjustments. This chapter looks into various approaches and suggestions for new menus during the environmental emergencies we know will come.

Perfect Storms

Several ecological crises are already on the way, unfolding rapidly due to climate change and human-driven transformations of the environment. These environmental disasters—such as biodiversity loss, freshwater scarcity, soil degradation, and a wetter and stormier climate—are increasingly threatening global food security.[3]

Massive food crises can also emerge suddenly due to large-scale catastrophes and black swan events.[4] Nuclear wars, asteroid impacts, pandemics, plant diseases, volcanic eruptions, and the end of phosphate,[5] to mention some of the many known risks (see the introduction), can suddenly disrupt global food supplies and cause rapid and severe food crises around the world.[6]

Extreme disasters on the scale of Fimbulwinter can even lead to the loss of elements that are vital in food production, such as sunlight. Large-scale nuclear warfare could inject massive amounts of dust particles into the stratosphere, blocking solar radiation for an extended amount of time. This would result in winter-like climate conditions, referred to as *nuclear winter*.[7]

A more recent Fimbulwinter-like crisis arose from the volcanic eruption of Mount Tambora in 1815. It resulted in a three-year volcanic winter that caused worldwide weather disasters, harsh climate conditions, and famine across Europe from 1816 to 1818.[8] It is not a question of if but when the next massive volcano eruption will happen. Most likely, it will be sooner than later. On any particular day, more than twenty volcanoes are actively spewing out the earth's innards.[9]

How are we to feed earth's 8 billion and growing global population if any of these extreme climate conditions disrupt the production and distribution of food? Or worse, in a doomsday scenario, what if several of these threats arrive simultaneously? We must transform our diet and prepare food production to withstand such extreme future scenarios. For now, how do we ensure food security during the ecological catastrophes already unfolding?

Stockpiling for Present and Future Crises

In Genesis, the first book in the Old Testament, Joseph warns Egypt that seven years of plenty will be followed by seven years of famine. To overcome the coming catastrophe, Joseph advises Pharoah to collect grain during the years of abundance for use in the following barren years.[10]

Likewise, food stockpiling has traditionally been the primary approach to ensuring a food supply for future emergencies. The food is stored before the catastrophe occurs.[11] This method has been applied by individual households and is the oldest food security technique used by state governments to prepare for unstable periods such as environmental disasters, wars, and conflicts.[12] With the expected increase in climate disasters, many governments have renewed interest in maintaining emergency food reserves.[13]

Stockpiling is also gaining increased popularity among individuals. After the economic crisis in 2008, there was a surge in the numbers of so-called preppers who actively prepared for future possible emergencies.[14] In 2020, the Covid-19 pandemic brought a new crisis and new insecurity for personal food supplies.[15] A severe situation tends to impact our eating habits. As the Covid-19 emergency has shown, we can expect new consumption patterns to linger even in the aftermath of a disaster.[16] Rethinking and recomposing our menus can also lead to a reevaluation of how we manage our food sources.

Eating Waste

Like the seven-year cycle between plenty and starvation faced by Egypt in the book of Genesis, the global community is projected to face increasing food crises due to changing environmental conditions. While the near future might bring large-scale food shortages, the current world is marked by the opposite—an abundance of food and large quantities of food waste. It is estimated that globally around a third of all food is lost or wasted. Consumers in Europe and North America trifle away about 95 to 115 kilograms of food per year. The estimated amount of food wasted globally is 1.3 billion tons every year.[17] Recent research indicates that this might be an underestimate, with the real number twice as high.[18] Since we produce food in such high abundance, could we, instead of wasting it, somehow stockpile it for the barren years to come?

In an address concerning the global issue of food waste, artist Tattfoo Tan described how he turns meal leftovers into a future survival supply in his project *New Earth Meals Ready to Eat* (*NEMRE*) (2012). The project was created in the aftermath of Hurricane Sandy, which hit the eastern United States in 2012. It caused significant and unexpected disturbances to food security. Local supermarkets were flooded, most products were thrown away, nothing was

replaced, and people were left with nowhere to buy goods. Tan suggests we should all brace for future crises by preparing today's waste for future eating. His *NEMRE* are accordingly made from food waste acquired from an undisclosed grocery store and then cooked, dried, and dehydrated in a package to last for up to a year. Seeing his project as a preparation for future crises we know will come, the artist encourages people to secure their own emergency food. In theory, by following this recipe, we could produce emergency provisions and stockpile nearly all of the 1.3 billion tonnes (1,000 kilograms per tonne) of global food waste every year.[19]

Critical questions remaining to be answered are the environmental costs of such a conservation endeavor. Significant amounts of energy would be needed to freeze-dry the food, transport it, store it, and create materials such as plastics for sealing bags.

While stockpiling is a viable solution to short-term disturbances in the food supply, it is also expensive and can disrupt precatastrophe food security.[20] It is hard to build up lasting provisions for more than a year. It makes even less sense in the case of a crisis that endures for several years. As David Charles Denkenberger and Joshua Pearce frame it: "Even in the best-prepared survivalist situation with a well-armed retreat and a complete mastery of wilderness survivalist skills, the survivalist starves to death along with everyone else when the food supply runs out after a year or two."[21]

Tan's project is nonetheless an interesting short-term food-supply solution in postcatastrophe situations. *NEMRE* also exemplifies the focus on decreasing food waste that has gained popularity in recent years. There has even been proposed a "World War on Waste."[22] This sounds like a sensible and ecological step, but for Denkenberger and Pearce, the reduction of food waste now will not necessarily help in a crisis and instead could make it far worse.[23] This seems like a paradox, but our current surplus of food will most likely act as a necessary and needed buffer should a major global crisis arise. Producing and eating only exactly what we need leaves us vulnerable to sudden shortages. Producing waste can be understood as a necessary surplus. A certain overproduction of food will give us the time and capacity to readjust and ramp up the production of alternative foods in times of crisis.

Figure 2.1. Tattfoo Tan, *New Earth Meal Ready to Eat (NEMRE)*, 2013. *NEMRE* boxes containing dehydrated food packages prepared using discarded food waste.

Alternative Agriculture

More extreme disasters, such as the interruption of sunlight, could affect the global supply of means for decades, adding to the unlikeness that stockpiles can feed the global population.[24] In such a situation, alternative agronomics is needed. Food security can be attained by adjusting the agricultural practices for food production in extreme conditions.[25] Such circumstances could be the long periods of a nuclear or volcanic winter. However, climate change, a reduced freshwater supply, and soil depletion have already drastically decreased the arable land per person.[26] Agricultural practices therefore need to be reinvented and adjusted now before the next crisis arrives. One suggested method for doing this is indoor vertical farming that does not require any soil and needs only a limited amount of water.[27]

In *Future Food Hack* (2015), artist Jimmy Tang explores various alternative approaches to agriculture. The project includes five do-it-yourself (DIY) maker kits envisioning new ways of growing food. One of the kits, called *Agara*, is designed for growing food without soil. It is an incubator that allows the users to grow plants from seeds using an agar growth medium. The kit demonstrates how food can be generated with minimal resources in small, enclosed environments. The *Agara* kit comes with several seeds and sterile test tubes that are used for germination and sprouting.[28] Another of Tang's kits, called *Skinseed*, speculates that food could be grown symbiotically through a wearable absorbent and seed-growing patch attached directly to the wearer's skin. The artist assumes that seeds will absorb the necessary moisture and heat from the user's body, in effect turning the body into a farming field. As further discussed in chapter 5, Human Deli, utilizing our bodies as farmland provides foods with the shortest supply chain possible. Short-traveled food also comes from the places we live, such as cities.

A range of approaches to alternative agriculture focuses directly on food production in the cities. From indoor vertical farming to rooftop gardening, urban agriculture is gaining popularity as a sustainable solution to future food supply.[29] With more than 68 percent of the world's population estimated to be living in urban areas by 2050,[30] urban farming makes sense. In a utopian

Figure 2.2. Jimmy Tang, *Future Food Hack*, 2015.
Agara kit for growing food in a soil-less environment.

Figure 2.3. Jimmy Tang, *Future Food Hack*, 2015.
Sprouted seeds inside the *Agara* kit plant incubator.

scenario, all the necessary food could be grown locally, eliminating the food miles that are usually extremely high for foods consumed in the cities.

In 2020, the world's largest urban rooftop farm opened in Paris. At 14,000 square meters, the big aeroponic city farm grows plants without soil, using only water and nutrients, and promises to produce up to 1,000 kilograms of fresh fruit and vegetables daily during the harvest season.[31] The project initiator Pascal Hardy believes that if enough urban spaces were used, it would be possible to produce between 5 to 10 percent of the food needed to satisfy the city's population.[32] Increasing vegetation in the cities will also combat global warming, especially the urban heat islands.[33] Paradoxically, as discussed in chapter 3, Anthropocene Specials, the urban heat islands contribute to creating a better microclimate for growing foods in the city.

For urban farming to contribute to food security in any meaningful way, it matters what foods are to be grown. Lettuce and strawberries grown on city rooftops have little to do with survival. Can rooftop gardening offer vital staple foods? Aeroponics, for example, is also proposed for cultivating staples like potatoes.[34] As soil and water become increasingly degraded in many areas worldwide, this could be beneficial.[35] With the current technologies, however, most of these approaches face significant challenges. Aquaponics—a closed-loop circular system for cultivating fish and plants—has an advantage of circular fertilization of the greens. On the downside, it is highly energy- and water-intensive.[36] The same issue relates to most of the high-tech agricultural approaches, including indoor farming, which requires artificial lighting. In the worst-case future scenario, however, it might be the only option, no matter the costs. Indoor farming can provide a highly controlled environment that avoids increasingly extreme weather events and yields harvests throughout the year.[37]

Other urban agriculture practices include insect farming (see chapter 6, Bug Buffet) and algae farming (see chapter 7, Future Superfoods). These ways of farming present innovations and advantages but also significant technological and food safety challenges.

Alternative ways of growing food and urban agriculture practices are viable as long as there is enough sunlight or artificial light for plants to grow. Would it be possible to produce food without sunlight?

Alternative Food Sources

In the most extreme future disaster scenarios, even the mentioned alternative agriculture methods might be rendered impossible. One approach left is the production of foods that do not need sunlight.[38] While alternative foods are the most viable approach for future food crises, they face the challenge of scaling up production and social acceptability.[39] Traditional examples include feeding cellulose or other indigestible plants to mushrooms and beetles that, in turn, are further fed to livestock or directly consumed by humans. Another approach is to utilize natural gas to feed bacteria that can also be fed to animals.[40] An extreme but unexplored possibility is using anaerobic chemosynthesis to produce organic material without oxygen and light.[41] This is done by several microorganisms in regions of the oceans devoid of light.

A hardly utilized resource is the massive amount of plankton in the oceans. This could become one of the new, green pastures for generations to come. Unfortunately, finding ways to harvest and turn plankton into human food on a massive scale is still as far away as the first and feeble attempts during World War II.[42]

Another untapped resource is the immense biomass of earthworms that are found all around the globe. Earthworms represent by far the largest wet biomass on earth. They were one of the favorite research topics of Charles Darwin. He was among the first to recognize their immense importance for the global ecosystem. Recent corrections to his early calculations estimate that earthworms have more biomass than the combined weight of all humans and farm animals.[43] Adding to this, earthworms are also among the best protein sources.[44] We literally have an abundance of near-complete meals crawling in the soil beneath our feet. Given the omnivore nature of humans, why do we not already eat them?

A primary challenge to the production and consumption of alternative foods is overcoming the many social and cultural norms in the Western diet connected to resources that are considered inedible or outright rotten.

Enjoying Rotten Food

According to Darwin's Origin of Species, *it is not the most intellectual of the species that survives; it is not the strongest that survives; but the species that survives is the one that is able best to adapt and adjust to the changing environment in which it finds itself.*
—Leon C. Megginson[45]

Future food-supply crises will push us to add new food sources to our diet. One way to do this is to look for entirely new foods to be harvested and eaten in the future. Another way is to explore ways to consume known and existing resources that are considered inedible. Humans already eat certain kinds of inedibles—that is, carefully rotted foods such as old cheeses, stinky tofu, hakarl (Icelandic fermented shark), and kimchi (Korean fermented vegetables). Culturally acquired tastes and existing food traditions play an essential role in determining the foods we consider edible and decide to eat. Rotten cheese and fermented shark meat are eaten by no one but those who have learned that these seemingly inedible products taste great. Normally our olfactory capabilities work as an early warning system telling us what foods are rancid and prone to cause serious digestive problems or even poisoning. Additionally, our diet is defined by the physical limitations of our bodies and our digestive system. There are foods we simply cannot digest. That could change.

The *Human Hyena* (2014) project by Paul Gong proposes a radical solution to future food crises through the expansion of our digestive abilities.[46] The project imagines how synthetic biology could transform the human digestive system to resemble that of hyenas. According to Gong, hyena species are unique eaters. They have an amazing sense of smell and taste, which, combined with their unusual digestive abilities, allows them to consume and digest rotten food, including rotten meat. Through the use of new synthetic bacteria and unique biotechnological tools, the *Human Hyena* project speculates on how this ability, when transferred to humans, could tackle the increasing scale of food waste and solve food shortages. Human hyenas would be able to turn the 1.3 billion tonnes of global food waste into a nutritiously precious and tasty resource at any stage of decay and without costly preservation.

Figure 2.4. Paul Gong, *Human Hyena*, 2014.
Hyena Inhaler.
Figure 2.5. Paul Gong, *Human Hyena*, 2014.

Gong has created a set of tools to help humans enhance and extend their diet. Through the *Hyena Inhaler*, the user can inhale genetically modified human and hyena bacteria that enable him to eat rotten food without becoming sick. Another tool in the project is the *Smell Transformer*. It uses the enzymes of a genetically modified miracle fruit (*Synsepalum dulcificum*) that turns any smell sweet. The same berry is also used in the *Taste Transformer*, making all rotten food taste sweet.

Indeed, if we could consume, digest, and even enjoy rotten food, including meat, even the most severe post-ecological crisis leftovers would satisfy our taste buds. The ability to digest food overdue its expiration date would also decrease the need for food preservation and related food safety regulations.

If food was considered safe to eat and even tasty when rotten, there would be less food wasted in the world. Even with disturbances in the food-supply chain, as discussed by Tattfoo Tan in his project *New Earth Meals Ready to Eat* (*NEMRE*), there would be no worries about food spoilage. Foods could be eaten long after the expiry date we consider mandatory to follow today.

In 2020, the Covid-19 pandemic led to significant disturbances in the global food supply. While many people were struggling to ensure they had enough to eat, food producers were also affected. In the United States alone, 1.5 billion pounds of potatoes were discarded as waste straight after harvest from the fields. There was similar waste with other vegetables, dairy products, and other items that today are considered perishable. While decay leads to the loss of vitamins and proteins in foods, farmers might be able to have more time to distribute the harvested goods.

What if we turn the focus entirely toward modifying our own bodies as a solution to feeding the world in the future? Could we, by demand, make anything remaining in the postecological catastrophes edible and taste great? Redesigning the human digestive system and diet opens up entirely new ways to imagine and shape the future cuisine.

Cellulose: A Staple for the Future

What is a waste for one species can be food for others. If we can transform our bodies to eat rotten food, can we also modify ourselves to digest substances

Figure 2.6. Gints Gabrans, *FOOOD*, 2014. Genetically modified metabolic bifidobacteria used to synthesize an enzyme that breaks down cellulose.

Figure 2.7. Gints Gabrans, *FOOOD*, 2014.

that traditionally do not hold any nutritional value for us? The earth offers many potentially consumable resources. Besides the new raw synthetic items of the Anthropocene, discussed in the next chapter (see chapter 3, Anthropocene Specials), we also have resources of natural origin that we have not yet learned to utilize as food.

During the food crises of the past, humans have turned to various items that otherwise would not be eaten. A traditional famine diet in Scandinavia, for example, included foods made using wood bark.[47] The Copenhagen-based organization Nordic Food Lab undertook culinary research of wood as an edible ingredient and created tree bark bread (2015). The lab investigated the Scandinavian practices of using the birch and pine bark flour in traditional Sami cuisine.[48]

The Nordic Food Lab made birch flour by collecting the inner bark (phloem) from vertical stripes cut into the tree, drying the bark, and finally grinding it into a powder using a mortar. Surprisingly, the inner bark contains 800 to 1,200 calories per kilogram that are digestible by humans.[49] Pine tree bark flour is made in the same way without the need to cut down the tree.[50] The bark's 800 to 1,200 calories per kilogram might seem low compared to sugar's 3,860 calories per kilogram, but it is still enough to qualify as emergency food.

While bread made from the inner bark is a famine staple, we gain little nutritional value from other parts of the tree. We cannot fully digest cellulose in the same way that some animals can, so many plants, including grass, are not part of our diet. Can we modify our digestive system to gain more from eating cellulose?

In the project *FOOOD* (2014), artist Gints Gabrans proposes genetically modifying metabolic bifidobacteria to give humans the ability to synthesize an enzyme that can break down cellulose. This would allow humans to fully utilize any cellulose-containing resource, including wood and even paper. Consuming wood would then give us three times as much energy as eating potatoes.[51]

Enhancing the human digestive system in the proposed way would allow us to gain the full nutritional potential of such foods as bark flour and maximize the energy acquired from all plants that are already part of our diet.

Emerging biotechnologies of the future might expand our diets immensely. Cellulose could become not only a staple food for future emergencies but also an important element in an everyday diet. Besides eating grass and trees from the backyard, we could even cook the furniture and eat books directly from the shelf. While humans cannot yet fully digest cellulose, the thought of eating books is playfully exercised in Land Rover's *In Case of Emergency: Eat This Book* (2012). The edible survival guide was launched by the Young & Rubicam Dubai advertising agency for the Land Rover car company in the United Arab Emirates. It included essential tips on how to stay alive in the Arabian Desert, and because it was made from edible paper, it could be eaten as a last-resort meal. The book claimed to hold nutritional value similar to that of a cheeseburger.[52]

In a perfect future scenario, synthetic biology would enable us to sustain ourselves on a cellulose-based diet full of trees, leaves, plants, and even an occasional book—as long as they might last. However, with the increasing scale of ecological crises, even possibly Fimbulwinter-like winters, we must also contemplate a future where cellulose is not as readily found as today.

Air Food

What other alternative food sources can we utilize when traditional agriculture is rendered impossible? In the darkest future scenarios, if animals and fish go extinct and soils deteriorate to a level where nothing can be grown anymore, what can we eat?

What are the fewest resources we need to ensure survival? What if we could make food the way we make fertilizer today—out of thin air? In the 1960s, the US National Aeronautics and Space Administration (NASA) investigated how to produce food for long-term space travel. One of its findings was that CO_2 could be used as the main ingredient to produce edible microbial biomass.[53] Microbial proteins, especially bacteria-based ones, are seen as a viable alternative for replacing the traditional animal- and plant-based proteins.[54] Dry bacteria-based biomass contains between 50 and 83 percent of microbial proteins. Another advantage is its high growth rate when compared to other protein sources.[55] To produce air food, we would not even need light, just water and energy.

At least two companies are currently taking the idea further and working toward turning air and energy into a commercially available product. A Finnish company called Solar Foods is developing a protein powder called Solein[56] using air, waste CO_2, water, and electricity. Solein contains around 65 percent protein, which is similar to that of soy or algae. The powder is said to be neutral in taste. But as discussed in chapter 4, Fake Foodies, even if future food sources are essentially tasteless, food science can make pretty much anything taste like anything else: bitter, sweet, sour, salty, or umami (the savory flavors of roast meat or soy sauce). As for now, the Solein powder can be used as an ingredient in traditional cuisine, in meat substitutes, or even as feed to grow meat in the lab.

Making food almost out of thin air would result in a guiltless protein without animal slaughter and without the huge environmental and carbon footprint that characterizes industrial farming practices. Air food might become not only carbon neutral but even carbon negative. If we were able to feast on the air in the future, what about spicing up our diet with solar energy?

Human Photosynthesis

As long as there is sunlight, we might also consider it as a possible direct food source. Could we eliminate the need for eating through the oral orifice and simply eat through our skin? What if we engineered ourselves to obtain the plant-centric skill of photosynthesis? Plants are commonly thought of as the only lifeforms to photosynthesize. There are examples of organisms in the animal kingdom, however, that indirectly also perform photosynthesis. The pea aphid (*Acyrthosiphon pisum*) is an insect that uses pigments to capture sunlight and transfer it within its cells for energy production.[57] And the sea slug *Elysia chlorotica* eats algal cells to enable photosynthesis. It can sufficiently sustain itself on the carbon energy for several months.[58] Could emerging biotechnologies enable us to develop similar dietary mechanisms? In the novel *The Possibility of an Island* by French author Michel Houellebecq (2005), future humans need only a little tablet of salts now and then, some water, and some sunlight.[59]

It is unlikely that human bodies will be able to perform photosynthesis in the near future. There are too many biotechnological challenges to overcome, including the relatively small surface-to-mass ratio of a human body to catch and harvest the necessary energy through sunlight,[60] which is even less if clothes are worn.

However, we could take a similar approach to the sea slug *Elysia chlorotica* and enter a symbiotic relationship with other nonhuman species that already perform photosynthesis. Such a scenario is investigated in the *Algaculture* (2011) project by Michael Burton and Michiko Nitta, where human organs are enhanced with algae, allowing humans to become semiphotosynthetic[61] (see chapter 7, Future Superfoods). Becoming symbiont with photosynthetic creatures such as algae or plant species would turn us into plantimals and require us to settle for more beneficial relationships with our environment.

Summary

The cuisine of the past is marked by the incredible human ability to source alternative foods in times of emergencies and extreme conditions. The knowledge of past meals will be useful when developing the ecological crisis menu of the future. However, previous time-tested famine staples will not be enough to ensure survival in this age of constant catastrophes. Even if a nuclear or volcanic winter does not happen in the next millennium, human-driven environmental changes are more than enough to severely disrupt global food security.

Stockpiling the existing surplus and the vast amount of global food waste can serve as a short-term solution to future crises. But we need new twists on emergency staples from the past. Our soon-to-be menu will be characterized by an alternative agriculture that could partly move directly into urban environments. While growing lettuce on rooftop gardens will not feed us sufficiently, closed-loop systems such as fully controlled indoor aeroponic and aquaponic farms could allow for the production of some nutritious staples. However, the products of alternative agriculture might come at high energy costs.

To survive, we will need to develop new foods and invent entirely new ways of eating. We could consume untapped existing resources such as plankton and earthworms and even find ways to utilize foods that are considered inedible. If bioengineering expands the capabilities of the human digestive system to digest cellulose, then this could become our primary source of energy. With similar methods, rotten food could appear to taste even more delicious than fresh food. And if trees vanish and all the rotten food we stockpiled is eaten, then we could use bacteria to make food out of air. A further speculative idea is to engineer our bodies into a symbiotic relationship with algae that would enable us to perform photosynthesis and feed on sunlight.

The massive geological and ecological transformations we have brought upon the earth will provide new nutritional foods to eat and previously unknown flavors to enjoy. The ecological crisis has arrived, and it does not mean we will have less food to eat. Just different. It is time to bring the Anthropocene specials to the table.

Anthropocene Specials

3   Anthropocene Specials

If you fail to prepare you are preparing to fail.
—H. K. Williams[1]

The Anthropocene diet will be defined by the ecological crises ahead. And while many foods of the future might simply be a twist on old emergency staples, this geological age will also bring entirely new ingredients and food experiences. Our diet has always been driven by cultural and social factors and by our ability to adapt to changing climate and environmental conditions.[2] Simultaneously, our eating habits and ever-increasing global demand for food have caused profound changes to the earth's biosphere.[3] Our dietary impact on earth is remarkable, and our food production and eating patterns (like an enormous consumption of meat and dairy) have accelerated the arrival of the Anthropocene.[4]

In the Age of Man, it becomes increasingly hard to separate the human-made from the natural. Animals like chickens and plant species like rice, soy, and the potato have gone through enormous transformations due to human interventions (see chapter 7, Future Superfoods).[5] Human activity and human-made synthetic materials have become inseparable parts of the environment, from microplastics present in air and water to increased levels of CO_2 in the atmosphere.[6] These changes bring along new material compositions, resources, and possibilities. While environmental changes present endless challenges to food security, the Anthropocene also comes with exciting possibilities for culinary explorations.

This chapter examines some new food opportunities that we can and will create for ourselves. How are human-made resources such as plastic, radiation, and pollution becoming part of and affecting the global food chain? How do they influence our taste and change our eating habits? And will these new ingredients become a standard element of future cuisine?

Chicken: Food of the Epoch

All periods have their styles of eating, their table manners,
their preferred dishes and styles of recipe writing.
—Timothy Morton[7]

Timothy Morton has said that all periods have their own styles of eating, but does this mean that all geological epochs have their own staple foods, too? It is impossible to generalize the diets of early human ancestors throughout the geological timescale.[8] However, the Pleistocene (from 2,588,000 to 11,700 years ago) is considered to have marked a critical shift toward a meat-based diet for early human ancestors and Neanderthals.[9] During the Pleistocene, it is assumed that modern humans living in Europe ate a significant amount of mammoth meat,[10] so mammoth meat might have been a revolutionary and epoch-defining staple food during that time frame.

A lot has changed since our ancestors hunted down and ate the mammoth to extinction two geological epochs ago (see chapter 9, Fantastic Cuisine). In the Anthropocene, our drive for ever bigger and tastier food has done something quite the opposite for one particular species. Humans have cultivated one single food so remarkable it is considered a geological marker of the Anthropocene. The domestic broiler chicken has become the most iconic food of our age.

The modern industrially farmed chicken originated from the red jungle fowl native to tropical Asia. From being a local, distinct bird with a relatively small population, its upgraded successors now represent the world's largest bird population.[11] About 25.9 billion chickens were estimated to inhabit the planet in 2019.[12] Its global population has bypassed every other bird species, including the passenger pigeon—the most common wild bird known to history, whose population peaked at 5 billion in the 1800s.[13]

Since the *Chicken-of-Tomorrow* program in the 1950s[14] that encouraged the development of higher meat-yielding birds, the domestic chicken has undergone

Figure 3.1. Nonhuman Nonsense, *Pink Chicken Project*, 2017.
Pink chicken with the future stratum of the Anthropocene.

Figure 3.2. Nonhuman Nonsense, *Pink Chicken Project*, 2017.
Eating pink chicken.

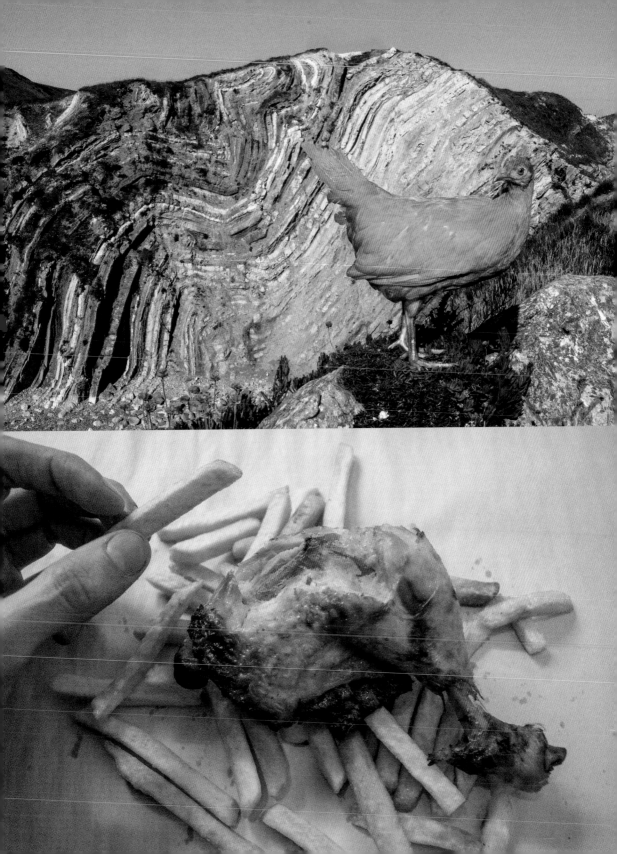

significant transformations. The two most notable contemporary features are the remarkably bigger skeleton and shorter growth time. The modern chicken goes from egg to fully grown and ready-to-be-eaten bird in five to seven weeks.[15] The broiler has been genetically modified to be bred mainly for meat. It is unlikely to survive longer than seven weeks, even if not slaughtered, due to its nonproportionally small organs such as the heart and lungs.[16] The broiler is here to serve its purpose in life—to be eaten by humans.

And we cannot get enough of its meat. Over 69 billion chickens were consumed in 2018 alone.[17] The enormous global chicken consumption leads to extreme amounts of carcasses discharged into the environment. Their massive numbers and distinctive physiological characteristics make chicken the key fossil species marking the earth's upper crust in the Anthropocene.[18] The intricate relationship between humans and chickens is well represented by artist Koen Vanmechelen in *The Cosmopolitan Chicken Project* (*CCP*) (1999). The project explores how humans affect the biological and cultural diversity of the planet through the lens of domesticated chickens.[19] The artist recognizes that chicken breeds around the world reflect the cultural characteristics of the community that has bred them. The ultimate goal of the *CCP* is to create a truly cosmopolitan chicken that carries the genes of all the chicken breeds from around the world.[20]

As a lasting reminder about the profound impact of human actions on this planet and the bird that grew to dominate the global diet, the *Pink Chicken Project* by Nonhuman Nonsense (2018) proposes to genetically modify the whole species of the domestic chicken (*Gallus gallus domesticus*) so that its color is pink. The project suggests using the genome editing tool CRISPR to permanently introduce a pink pigment into the bird's bones and feathers. Nonhuman Nonsense speculates that by using the genetic system—a gene drive—it would be possible to forever alter the entire global chicken population in just a few years. The introduced pink pigment would be fossilized together with the calcium of the bone. Then the pink layers of chicken bones would form a geological imprint visible to the naked eye during future excavations in the earth's crust. The pink line would additionally testify to our role as successful biohackers and breeders of chickens and the important role the chicken played in terraforming earth. Nonhuman Nonsense further proposes encoding the pink chicken DNA with a message for the future, demanding an action to stop the sixth mass extinction of species.[21]

Chicken bones will continue to pile up as long this bird is farmed and valued as food. It can no longer survive without humans. Ironically, an abrupt halt to the pink chicken line in the crust would most likely signify that a dreaded global mass extinction was completed.

Tasting Extinction

While the domestic chicken has risen to a dominating position in the Anthropocene, many other animal, plant, and insect species are going extinct. The earth's habitat has always been in a constant change marked by evolution, adoption, interactions between different species—and extinction. According to scientists, the sixth mass extinction is well on its way. An event of mass extinction is defined as when "Earth loses more than three-quarters of its species in a geologically short interval."[22] The current trajectory of biodiversity loss on earth is widely linked to the results of human activity. What is also known as Anthropocene extinction or the Holocene extinction could reach a scale similar to the previous five mass extinction events within just a few centuries.[23] These massive changes to the earth's biodiversity mean we will be taking some of our favorite dishes off the menu.

Humans have a long history of eating whole species to death. Such was the likely demise of meaty mammoths, which were driven to extinction due to climate change and increased human pressure through intensified hunting.[24] Europeans first encountered the Steller's sea cow in 1741, and it was entirely exterminated by the year 1768[25] due to intensive and wasteful slaughter by Russian hunters with primitive hunting technologies.[26]

The most famous and well-studied culinary extinction is the passenger pigeon in North America. The case is closely examined by writer Lenore Newman in her book *Lost Feast: Culinary Extinction and the Future of Food* (2019).[27] She describes passenger pigeons as once having a population of as many as 5 billion birds. The massive flocks of this migratory and tasty bird, which are said to have blocked the sun for hours or even days, were the reason Americans thought they could never deplete their resources. With the invention of the telegraph and railroad, the passenger pigeon was unsustainably harvested to death. As many as 650 birds would be taken down with a single net. Thousands were fed to pigs and even used as fertilizer. With the declining population, the passenger pigeon turned into America's upper-class delicacy until it became

extinct in the wild by the 1880s. The last specimen, Martha, who lived in the Cincinnati Zoo, died in captivity in 1914.[28]

Today many species—including the Atlantic bluefin tuna and many shark species that are being eaten to the brink of extinction—will be in the human diet for only a short while longer. While Atlantic bluefin tuna is valuable because of its delicious meat, sharks' fins are used to make a traditional Chinese dish, shark fin soup, which is made at a high environmental cost.[29] This food item is often harvested through the practice of shark finning—cutting off the fins and often discarding the rest of the still-living shark carcass back to the ocean.[30] It is a delicacy many are willing to acquire, even if illegally. And humans continue the unsustainable killing even if the inevitable consequence of extinction is well understood. Some species, such as recently found birds from the Brazilian coastal forests, were lost even before they were discovered by humans.[31] Species have disappeared from the earth's surface even before humans have had a chance to taste them.

To reflect on a future cuisine marked by food scarcity and biodiversity loss, Miriam Simun and Miriam Songster developed *GhostFood* (2013). In the project, the artists pair synthetic scents and facsimile edible ingredients to simulate the taste experience of foods that soon might become unavailable due to climate change. These food simulations are achieved using a wearable device that delivers scents under the nose, allowing for direct olfactory stimulation.[32] The *GhostFood* menu offers endangered foods such as Arizona peanut butter, chocolate made from cocoa beans from West Africa, and Atlantic cod dishes. In the 1990s, the latter was fished to near extinction.[33]

The project is a visceral reminder that the consequence of the sixth mass extinction is a loss of the familiar. Many more of our favorite foods might soon live only in our memories or be brought back synthetically. In the future, we might be able to bring an extinct species back to life through the use of biotechnologies, just like speculated in the *Jurassic Park* movie franchise (see chapter 9, Fantastic Cuisine). With de-extinction technologies still in their infancy, what new Anthropocene specials are already available?

Figure 3.3. Miriam Simun and Miriam Songster, *GhostFood*, 2013. The smell-emitting device.

Figure 3.4. Miriam Simun and Miriam Songster, *GhostFood*, 2013. The *GhostFood* trailer.

Invasive Cuisine

While humans care for species they see as rightful inhabitants of their surroundings, others are perceived as a threat that needs elimination. Such is the case of invasive species—non-native organisms that adapt to new geographies and locations and spread aggressively. Invasive species often cause adverse environmental effects and threaten native biodiversity, economies, and even human health.[34] They also can be seen as filling a unique gastronomic niche—the invasive diet. Harvesting invasive species for consumption often is suggested as a way to limit their spread.[35] Could we really eat our problems away?

Human success in driving countless animal and plant species to extinction tells us that we can do the same with invasive species.[36] For example, Norway's southern coasts are increasingly taken over by the Pacific oyster (*Crassostrea gigas*) from the Pacific Northwest near Japan. These oysters are rapidly changing the local habitat, pushing native species to extinction, and occupying and ruining beaches and ecosystems. Yet they provide a nutritious, tasty, luxurious delicacy in abundance. They are now so plentiful that the authorities are asking people to catch and eat as many as they can, preferably into extinction. Also, migrating populations of jellyfish can be an excellent source for healthy future superfood snacks. Fried, they taste delicious. With the same approach, invasive insect populations and insect pests might be controlled through harvesting for food (see chapter 6, Bug Buffet).

However, an invasive diet can culminate with paradoxical results. When invasive species are promoted as food, a new marketplace emerges that can positively affect the local economy.[37] The initially feared species can become an important local food source and even gain cultural value and attachment despite the potential damage it does to the local ecology.[38] There have been cases of locals protecting the invasives from eradication, reintroducing them if eradicated, and purposely spreading their populations to other areas to gain economic benefits.[39]

If an invasive species, such as the American signal crayfish, is more successful than species native to Europe, why eradicate them? It is a thriving food source that tastes just as good and is bigger than the European freshwater crayfish species. Another example is the invasive king crab in northern Norway, which now has become a valuable catch for fishermen.

The Anthropocene has arrived with an update to local menus around the world. While invasive species are new food sources of organic origin, other ingredients soon to be on our plates will be purely synthetic. They are changing our diet while we hardly notice it.

Radiation-Bred Food

Not all the new Anthropocene resources are visible to the naked eye. Some are hyperobjects—entities of such vast temporal and spatial dimensions that they are unfathomable.[40] Human-made radiation is one of the hyperobjects that have become part of our environment.

As an effect of the atomic age, all food grown after World War II is radioactively imprinted. The first nuclear weapon testing in 1945 in Mexico marked the emergence of human-made radiation on earth. More than five hundred nuclear tests have been conducted since then, releasing a considerable number of radioactive isotopes, including cesium-137, into the atmosphere, soil, and oceans.[41]

Many foods on the earth contain naturally occurring radioactive isotopes, such as potassium-40 in bananas[42] and radium in Brazilian nuts.[43] The nuclear radionuclides, however, do not occur naturally on earth. Therefore, nuclear radiation has been proposed as yet another possible geological marker of the Anthropocene.[44]

While there are different types of human-made radiation, nuclear radiation can quickly spread through the environment, including our food system, leaving long-term traces. The Fukushima nuclear disaster in 2011 led to the release of radionuclides in the North Pacific Ocean, enabling the spread of radioactive material through seafood. After the accident, the Pacific bluefin tuna transported Fukushima-derived radiation all the way from Japan to California.[45]

Radiation and its environmental effects are long-lasting. Mushrooms in some Czech Republic forests still contain nuclear radionuclides thirty years after the Chernobyl nuclear disaster in Ukraine.[46] It is even possible to verify vintage wines by measuring the amounts of cesium-137 through unopened bottles. Grapes grown before the first nuclear tests in 1952 are cesium-137 free. At the same time, the radiation levels for wines bottled between 1952 and 1980 vary year to year, helping to validate if the wine content matches what is written on the label.[47] Exposure to large quantities of human-made radionuclides is dangerous and carries several potential health risks. Luckily for many, the radiation level in vintage wines after 1952 is rather low and therefore within food safety regulations.

If we are so unaware of both the natural and human-made radiation present in various foods we eat, what is its actual role in the modern food system? Can it be tasted or even beneficial in food production?

While nuclear radiation has unintendedly become part of our diet as a result of nuclear fallouts, humans have been intentionally using radiation to enhance taste and modify crops since the 1930s.[48] Through a process called *radiation breeding* or *mutation breeding*, several crops (including wheat, rice, barley, and beans) have been improved using X-ray and gamma-ray radiation.[49]

Inspired by the emerging technology of radiation breeding, the futurist Filippo Tommaso Marinetti suggested in his "Manifesto of Futurist Cooking" (1932) that Italian cuisine should be upgraded with a variety of scientific instruments that apply radiation to foods. His concept of ozonizers would "give the liquids and foods the perfume of ozone," while ultraviolet ray lamps would activate new properties such as making foods more assimilable.[50]

How far has radiation breeding developed since the 1930s? To draw attention to the high number of radiation-bred foods in the modern diet, The Center for Genomic Gastronomy developed *Cobalt 60 Sauce* (2013). *Cobalt 60 Sauce* is a barbeque sauce made from five commercially available mutation-bred ingredients. According to The Center for Genomic Gastronomy, currently, 2,500 mutant crop varieties are registered with the United Nations (UN) and the International Atomic Energy Agency (IAEA). Many of which are commercially available in predominantly American supermarkets. These also include the plants used in the *Cobalt 60 Sauce*: "Rio Red Grapefruit, Milns Golden Promise Barley, Todd's Mitcham Peppermint, Calrose 76 Rice and Soy."[51]

Radiation breeding has become an essential technique in food production. Yet due to the associated dangers of nuclear radiation, its use and public acceptance are still limited.

New Edible Resources

Human activity on earth brings entirely new possibilities for eating and dining. The domestic chicken will in time become a geological indicator of the new age because of its massive presence as a staple food. Meanwhile, new raw human-made materials are still finding their place in nature and in the global food

Figure 3.5. The Center for Genomic Gastronomy, *Cobalt 60 Sauce*, 2013.

Figure 3.6. The Center for Genomic Gastronomy, *Cobalt 60 Sauce*, 2013.

TIMELINE

1896	1939	1945	1957	1958	1966	1971	1977	1984	1962–2010
RADIOACTIVITY DISCOVERED	USA STARTS ATOMIC ENERGY PROGRAM	NUCLEAR BOMBS ARE DETONATED IN JAPAN	INTERNATIONAL ATOMIC ENERGY AGENCY (IAEA) FOUNDED	GAMMA GARDENS CREATED IN NY FOR MUTATION BREEDING	MILNS GOLDEN PROMISE BARLEY CREATED	TODD'S MITCHAM PEPPERMINT CREATED	CALROSE 76 RICE CREATED	RIO RED GRAPEFRUIT CREATED	THERE ARE 172 MUTATION-BRED VARIETIES OF SOY REGISTERED ON THE IAEA DATABASE.

YOU CAN SEARCH FOR OTHER MUTATION-BRED PLANTS ON THE DATABASE: HTTP://MVGS.IAEA.ORG/SEARCH.ASPX

chain. While radiation can be employed to enhance already existing foods, what edible ingredients, tastes, and food experiences are we yet to discover as a beneficial consequence of the Age of Man? What will become the iconic staple foods of the future, made possible because of the Anthropocene?

In Timothy Morton's universe of thinking about dark ecology, in addition to radiation, other hyperobjects are plastic bags, global warming, and pollution.[52] This indicates that ecological thinking requires us to include these objects into the global food equation. The perhaps most symbolic hyperobject of the Anthropocene is plastic. What happens when this new resource enters our foodscapes? Can it be transformed into an edible product and even be appreciated by the human gustatory system—our sense of taste?

Plastics on the Menu

Alongside chicken bones and radiation, plastics are the third proposed mega-geomarker of the Anthropocene. Plastics have been produced in enormous quantities since 1950s.[53] Plastic fabrication has risen from 2 million tonnes in 1950 to over 300 million tonnes produced annually today. By 2017, altogether, 8.3 billion tonnes of plastic had been made globally.[54]

Just like radiation, plastic has been part of the environment for decades. Even if we were to halt all plastic production today, it would likely remain in the environment for an unforeseeable amount of time. Plastic interacts with nature in new, unpredicted ways. It has even formed a new type of rock, Plastiglomerate, when encountering molten lava.[55] While plastics are often used to produce larger objects, with time, they splinter into tiny pieces until microplastics—plastic particles smaller than five millimeters—are formed. They can further split into less than one-millimeter-wide nanoplastics that are invisible to the naked eye.[56] Because this human-made material is insoluble in water and resistant to biological decay or chemical attack, plastic remnants last for decades to centuries in the environment[57] and have now become inseparable from what we call *nature.*[58]

While plastic has become an essential element in food transportation, preservation, and safety,[59] we never intended to eat it. According to projections, there will be more plastic than fish in the world's oceans by 2050.[60] This is now a known and massive environmental problem. It is also a material that unavoidably ends up on our plates. Plastic food packaging can smell and even taste like food. Therefore, fish not only eat it by accident but regard it as a sort of junk food.[61] Thus, plastic waste in the oceans is turning into seafood as it is accumulated in fish. Given the rather large role fish plays in the human diet, it is now the primary source of plastic through the foods we eat.[62] Plastic has also been identified in sea salt,[63] honey,[64] and even beer.[65] Besides many other food sources, there is evidence of plastic in the air[66] and in tap and bottled drinking water.[67] By now, we eat, drink, and even breathe plastic.

The effects of microplastic particles on human health can be many. Accumulated levels of plastics containing polychlorinated biphenyls (PCBs) and other chemicals are linked to harmful health effects such as cancer, weakened immune system, and reproductive problems.[68]

Plastic might seem like a waste nature has no use for. While humans, fish, and animals are unable to digest it or gain any nutritional value, it does not mean that no organism can. In recent years, several studies have reported on organisms that can digest plastic.[69] Waxworms, also known as Indian meal moths, were discovered to eat polyethylene plastics.[70] Also, the bacteria *Ideonella sakaiensis* 201-F6 can use polyethylene terephthalate (PET) as an energy and carbon source.[71] With this knowledge, our plastic problem could be turned into feed for other organisms. With all the excess plastic in the environment, it might become a significant, perhaps even advantageous, part of the global food chain.

LIVIN Studio explores such a scenario in the work *Fungi Mutarium* (2014), created in collaboration with Utrecht University in the Netherlands. *Fungi Mutarium* is a prototype of a kitchen device that grows edible fungal biomass on plastic waste. The project uses fungi named *Schizophyllum commune* and *Pleurotus ostreatus* that are able to degrade toxic waste materials, including plastics.[72] The fungi are cultivated in bowls of agar—a seaweed gelatin substitute that acts as a nutrient base. They then digest the plastic fed to them and produce edible fungal biomass, mainly mycelium. The designers are calling these objects *FUs*. And they are ready to be eaten in only a couple of weeks.

LIVIN Studio argues that food production must be revolutionized with the use of new technologies due to the increased need for food farmed under extreme environmental conditions. To speculate on a culinary shift toward a future where we can eat plastic, LIVIN Studio has also created unique tools called *Fungi Cutlery* to prepare and consume *FUs*.

With increasing threats to our global food security, plastic might become a useful raw material of abundance in nature. In the future, we might forage

the small lost plastic pallets from the beaches, also known as *mermaid's tears*, as a nutrient to produce edible *FUs*. Plastic picking for food might become the mushroom and berry picking of the Anthropocene.

The enormous quantities of plastics on a global scale make the material a hyperobject beyond comprehension. Despite its undeniable presence, we do not want to consider it a part of nature. Instead, it is something we want to hide. However, we cannot pretend it does not exist anymore. And why should we if plastic can give us up to 300 million tons of edibles every year? This massive amount of potential nutrients could contribute to future food security.

Consuming Pollution

Using plastic waste as a feed to organisms that in turn become food for humans is a promising scenario. What other types of potentially edible ingredients can be recovered from waste and pollution? Can air and water pollution benefit our diet and offer new culinary experiences? Our cuisine has always been marked by the use of different spices and taste enriching ingredients. Can we find flavors unique to the Anthropocene?

Smog Tasting (2011) by The Center for Genomic Gastronomy does exactly this. The project allows us to taste air pollution. Through the use of egg foams, smog is harvested from different and highly polluted areas. As egg foams are formed of up to 90 percent air, the whipping process allows air pollution particles to be trapped in the dough.[73] According to the artists, the outcome is an assortment of biscuits that provide a variety of smog flavors based on the air we breathe every day. While the project is created as an awareness-raising experiment to address high levels of air pollution, it also contributes to defining and capturing the tastes exclusive to the Anthropocene. Air pollution is something many people are exposed to every day. Therefore, it can become a norm and feel natural.

Human-made pollution can affect food production in many unforeseeable ways. It can even improve it. Anthropogenic climate change in urban environments such as increased temperatures resulting in urban heat islands is seen as a phenomenon that needs to be reversed. To mitigate the effects, some have proposed popularizing urban farming.[74] Paradoxically, the unique microclimate of the urban settings could potentially provide a climate for growing plants that

Figure 3.7. LIVIN Studio, *Fungi Mutarium*, 2014.

Figure 3.8. LIVIN Studio, *Fungi Mutarium*, 2014.
Growth sphere with agar *FUs*.

Figure 3.9. The Center for Genomic Gastronomy, *Smog Tasting*, 2011. *Smog Tasting* meringues made by whipping egg foams in India.

cannot grow in rural areas just outside the city.[75] Anthropogenic pollution, such as exhausted heat from buildings and cars, might actually be improving the microclimates in cities, resulting in longer and hotter growing seasons and warming winters.[76]

How does pollution alter the senses of taste and smell? And can we actually taste air pollution? The new Anthropocene flavors, like air pollution, might one day even become desirable. Could they be consumed not only to enhance the flavor of foods but also to provide health benefits? If we use herbs because of their natural healing properties, can we extract similar elements from the polluted environment?

To explore the potentially valuable ingredients we waste through water pollution, Jon Cohrs and Morgan Levy created *Alviso's Medicinal All-Salt* (2010). It is a unique, hand-harvested salt enriched with pharmaceuticals from the waters of the Artesian Slough, a salt-water inlet bordering the Alviso district in San Jose, California.[77]

The pharmaceutically rich components come from Silicon Valley's local wastewater treatment plant. Wastewater treatment plants filter out most toxic contaminants but not the pharmaceuticals that many flush down the toilet. Valuable drug compounds such as antibiotics and antidepressants can therefore be reharvested through the recycling of wastewaters. The rich composition of *Alviso's Medicinal All-Salt* makes it a multipurpose product. It can be used for anything from salting your steak to treating depression and flu.

The artistic approach here is highly discursive. While locally sourced and recycled wastewater contains a complex cocktail of chemicals that are harmful to human health, the project speculates on all the useful ingredients that are wasted during the recycling process. In effect, the project contributes to a model of circular waste economy and recycling of resources that can impact and even benefit our future cuisine.

Summary

Until recently, our diet has been characterized only by foods of natural origin. However, in the Anthropocene, many animals and plants are genetically engineered to increase harvests and simplify the food production process. Breeding and modifying the broiler chicken to meet our insatiable appetite for meat are

Figure 3.10. Jon Cohrs and Morgan Levy, *Alviso's Medicinal All-Salt*, 2010. Pharmaceuticals-rich salt that is distilled from effluent wastewater.

likely to turn the massive deposit of chicken bones into a geological marker in the earth's crust. Unlike the domestic chicken, many animal, plant, and insect species are going extinct. The unfolding Anthropocene extinction is likely to take some of our favorite dishes off the menu in the near future. To keep the extinct foods in our diet, the *GhostFood* project suggests simulating their taste by pairing synthetic scents and facsimile edible ingredients.

Then again, there are plenty of new organic resources, including invasive species, that we can add to local menus around the world. While the invasive diet suggests the elimination of these non-natives from the environment entirely by harvesting them for consumption, their presence can boost the local economy and diet. Simultaneously, new human-made raw materials on this planet—such as plastic, radiation, pharmaceuticals, and air pollution—enter and gradually become part of our environment and also food systems. While initially ending up in our stomachs by accident, in the future, we might gladly include these ingredients in our diet, making them regularly appear on our plates. And against contemporary projections, their presence in our food system can contribute significantly to both our food security and new culinary experiences. Human-made radiation has already been used intentionally in mutation breeding to improve crops. Plastics can become a nutritious feed for organisms like meal moths, fungi, and bacteria, which can be further turned into nutritious edibles. Air pollution can spice up foods and capture the true taste of the Anthropocene, while polluted wastewaters hold valuable compounds, including antibiotics and antidepressants, that can be reharvested.

Finding a use for new raw ingredients is a compelling way forward toward adaptation and survival in the Anthropocene. The Anthropocene specials make us both question and expand our omnivorous diet. How can we upgrade and even exchange our traditional foods with the novel ingredients that are about to arrive? Including the new and strange is a good start. These are ample reasons to rethink the differences between foods and nonfoods. Can we tell the so-called real and natural apart from the fake and artificial? And will the differences even matter in the future, as long as these new foods give us the necessary nutrition and taste good?

Fake Foodies

4 Fake Foodies

The new soon-to-be edible raw ingredients of the Anthropocene, such as plastic and genetically engineered foods, can feel far too unfamiliar and strange. How can we change our dietary patterns when traditional dishes are hard to give up? Luckily, humans are masters of disguise and imitation when it comes to creating food. What happens when we hide the new and unknown in the skin of the old and familiar? This chapter explores the potential of using synthetic and alternative organic substitutes to replace the main ingredients of the original foods. Here, the term *fake foods* refers to edible items that mimic the taste, smell, and looks of a real dish. The nutritional value might be the same or better, but the actual contents—somewhat unexpected.

Fake food is a controversial topic. The thought that we might be tricked into eating imposters without realizing it evokes panic and shock. A fear of the fake has led to many fascinating food legends and hystories—the cultural phenomenon of "individual hysterias connecting with modern social movements to produce psychological epidemics."[1] Several controversial fake food stories have originated in Asia. From artificial eggs made from synthetic resin to rice produced from plastic, several videos and news articles have circulated the internet, warning about the fakes.[2] The main argument in these faux-food stories is that replacing the original ingredients with alternative substitutes results in cheaper production costs, benefiting the producer or seller.[3]

Between 2011 and 2016, stories about fake Chinese rice made from potatoes mixed with synthetic resin spread across Asia and some parts of Africa. The written evidence was supplemented with videos on how to identify fabricated rice by a burning test. The public panic led to an extensive investigation by major international food and safety authorities. Although serious food scandals—such as the adding of melamine, a chemical used to make plastic, to infant milk powder in China in 2008—have raised awareness about food safety,[4] no evidence was found that fake rice or fake potatoes were sold in the market. It is believed that the fake rice rumors originated in an attempt to protect local industries and raise public distrust of rice imported from China.[5]

Similarly, in Bangladesh, rumors of artificial eggs imported from China led to public outrage and confusion about the food safety of eggs.[6] Materials online even attempted to analyze the ingredients and procedures of producing fake eggs, claiming that the process used to make the parts of the artificial egg (yolk, white, shell) used chemicals that are harmful to human health.[7] The panic triggered a scientific study analyzing almost four thousand eggs collected from ports across Bangladesh between 2017 and 2018. No evidence of artificial eggs was found.[8] Looking at the long list of chemicals supposedly used in the fabrication, it becomes evident that it is much cheaper to let the chickens do the job than to undergo the labor-intensive preparation of faux eggs.

We are unlikely to be eating fake eggs for breakfast in the near future, but designer Matt Brown in his project *Food and the Future of It* (2010) imagines a futuristic food preparation device that would allow us to directly print eggs from our kitchens. *Egg Printer* is supposed to grow and cook eggs that look real from the outside but have fully customizable inner patterns.[9] The project proposes that if we could artificially replicate the nutrition value, taste, and texture of traditional food, we would be likely to imitate the looks too. We like the familiar and find comfort in habits when it comes to food, which is all the more reason that wild and uncanny food stories resonate in us.

Urban legends about fake foods have been around for a long time. A well-known fake food story was published in the *Washington Post* in 1962. The article warned that Italian food fakers were making "Parmesan" cheese from various waste materials, including the plastic binders used to produce plastic umbrella handles.[10] The fake product supposedly had been on the market since 1957, and surprisingly, there had been no complaints from customers.[11] Since then, the story has lived on and been mentioned in several books as a legendary case in which "a product sold as grated Parmesan actually consisted of grated umbrella handles."[12]

Although never validated as true or false, the debate about the contents of grated cheeses has repeatedly reappeared in the public eye. And even if no cheese was ever made by manually grating umbrella handles, in 2016 it was reported that cellulose from wood pulp is frequently used as a natural anti-clumping agent for pregrated cheeses by several brands in the United States.[13]

Another interesting case is the Chinese wine counterfeiting industry. Wine drinking in China can be considered a relatively new cultural phenomenon.[14] Since the 1990s, the wine market has grown rapidly, allowing the wine industry to thrive. However, it is estimated that in 2013 over 70 percent of wine brands sold in China were not authentic and, in some cases, were not even wine. Counterfeit wines can be made cheaply and easily from a mixture of water, wine juice or edible alcohol, colorant, fragrant essences, and other chemical ingredients, including thickening agents and preservatives.[15] The final product resembles wine, but in many cases, the chemical ingredients used pose a serious health risk. These fake wines are sold either by reusing and refilling recycled bottles or faking original labels. Many if not most of China's wines are simply knockoffs. China does not have a long tradition of wine drinking or production; therefore, the Chinese consumers simply lack the wine knowledge to tell the fake apart.[16] There even have been reports of Chinese customers who complain about fake wines only to reveal that they were so accustomed to counterfeit products that they thought the real was fake. If you have never drunk real wine, how can you tell it is real?

If our taste preferences can be culturally programmed to appreciate the fake as being real and if we occasionally consume wood pulp on top of our spaghetti without noting the difference, how many other foods could be altered or outright produced utilizing filler or replacement ingredients? What if replacement products could reduce the price of the original or even improve its quality? Are they then fake? What benefits are there for producing and eating fake foods in times of ecological crises? If we have so many foods in our diet that are unsustainable to produce using traditional methods, could fakes become more sustainable to produce?

More often than not, fake foods have negative reputations, and their potential health hazards help to keep them taboo. Several fake food concerns have turned out to be hoaxes. Nonetheless, these food tales stimulate the debate about what food can be. What if fake food is the answer to our soaring food security concerns, after all? Perhaps making eggs from polymer resin and rice from plastic are ideas worth further investigation for a hungry world that, at the same time, drowns in plastic waste.

Fake Food Futures

To satisfy the growing global consumption of food, it is projected that in 2050 we will need to produce 70 percent more food.[17] Most of the demand will be for livestock products. It is estimated that in 2050 the world will consume 70 percent more meat and 80 percent more dairy than in 2015.[18] How are we to ensure global food security for an increasingly hungry world if we are already using around 80 percent of all agricultural land to grow livestock?[19] Agriculture already has a significant negative environmental impact by reducing biodiversity, accounting for a quarter of all global greenhouse gas emissions, and polluting terrestrial, marine, and freshwater ecosystems through the use of fertilizer.[20] In the future, it will be impossible to sustain the global population on diets heavily dependent on traditional agricultural practices and industrial livestock farming. If conventional food sources will no longer be a viable option, how can we keep our favorite foods on the menu? Fake foods might satisfy our sentimental desire for foods that will no longer be sustainable to produce or ethical to consume.

If the answer to our future food security is fake food, under what conditions are we willing to eat it? The fake and artificially produced food should be better, tastier, and more environmentally friendly than real food. Who would say no to that?

Imposter Foods

With the massive environmental impact of industrial livestock production, animal-based products such as meat and dairy have become a major focus area in the search for substitute foods.[21] Vegetarian and vegan dietary trends, especially among young people in high-income countries, are further pushing the food industry to look for alternatives to animal-based protein.[22]

Most fake-meat products currently available commercially are created out of organic food substitutes such as tofu, seitan, and all the possible vegetables and nuts.[23] The market strategy is to mimic the original food perfectly. The end products are burgers, sausages, and turkeys that resemble the old meat classics. The ultimate goal is to fool food critics in a blind taste test to show that the fake products are like the originals in looks, taste, smell, and even nutritional value. The brands use conventional food terms mimicking the original dishes—*vegetarian burger*, *vegan bacon*, *almond milk*, and *tofurky*. There have been heated debates in Europe about whether using these terms should be banned for products that are not of animal origin. In 2017, the European Parliament banned the use of the words *milk, cream, cheese, butter,* and *yogurt* for plant-based products.[24] This legislation was almost extended to faux-meat products; however, the European Parliament rejected the proposal in 2020.[25] You can still find a veggie burger on the menu, and it has never tasted better.

The plant-based faux-meat industry has grown rapidly in recent years, and products keep improving. In 2016, Impossible Foods launched its signature product—the Impossible Burger. While all earlier versions of vegan burgers lacked a meaty taste, the Impossible Burger was celebrated for cracking the code and making its product taste like real meat. The heme protein gives animal meat its color and taste and is usually lacking in faux-meat products. Scientists from Impossible Foods discovered that heme can be extracted from plants. The company extracted the heme protein from the soybean plant and inserted it into a genetically engineered yeast, which allowed plant-based heme to be made in large quantities, giving every Impossible Burger its meaty taste and looks.[26]

Even without the genetically engineered heme protein, companies have found various other ways to manufacture their products to be as close to animal meat as possible. But with increasingly better results of meat mimicry emerge new dilemmas. Not everyone wants the fake meat to taste like real meat. In her book *Lost Feast: Culinary Extinction and the Future of Food*, Lenore Newman describes an experience of preparing plant-based burgers for two friends with different dietary preferences—one vegetarian and the other carnivore. The vegetarian burgers were made by the company Beyond Meat, another popular meat replacement in the market that also was released in 2016. The feedback from her dinner guests was unexpected. The meat lover was almost satisfied with the taste, but the vegetarian was slightly disturbed: "the more I chew it, the meaty taste becomes kind of overwhelming. It's just too close, too . . . I mean, it isn't bad. But I find it a bit disturbing. I think I'll stick with the yams."[27] As with the Chinese consumers who became used to drinking fake wine, a true vegetarian found that the fake burger resembled meat too closely. The fake was simply too real.

While meat eating has been an important part of our evolution and history as the human species, not all nonmeat eaters would like their vegetable patties to taste like meat. Replacing meat in our diet also means much more than just changing the source of the protein and making it taste right. How will our food traditions change if the future is meatless?

Research-led designer Hanan Alkouh created the project *Sea-Meat Seaweed* (2016) to explore whether we can replicate not only a meat-eating culture but also the traditions of meat keeping in a post-meat world. To address concerns about the meat market's massive role in environmental problems, such as the enormous CO_2 emissions from livestock production, the project uses a natural and organic alternative to meat—the red marine alga called *dulse* (*Palmaria palmata*) that is widely available along the Pacific and Atlantic coastlines. This alga is a perfect match for replicating meat due to its unique flavor. When fried, it tastes like bacon.[28]

The *Sea-Meat Seaweed* project emphasizes keeping and yet transforming the ancient and rich rituals of meat eating. Although Alkouh is against meat eating, the project recognizes it as a significant part of human society and culture. The project therefore replicates not only the looks of meat but also an infrastructure of a meat-free industry that mimics the roles of the farmer, slaughterer, and butcher. The project raises crucial questions about the value of the collective social and historical behaviors that will be eradicated in a post-meat world where animal farming is no longer socially acceptable. What if new foods significantly and possibly negatively impact the fabric of society's norms and values? Although the question here remains unanswered, the artistic importance of the project lingers in raising such intricate dilemmas.

Even if seaweed is considered one of the future superfoods and is a viable alternative for animal protein in terms of nutritional value, what if humans' carnivorous nature means that not everyone will be satisfied with plant-based meat replacements?

Lab-Grown Meat

In his article "Fifty Years Hence" (1931), Winston Churchill remarkably wrote: "We shall escape the absurdity of growing a whole chicken in order to eat the breast or wing, by growing these parts separately under a suitable medium."[29] Almost a century ago, this idea was thought well ahead of its time and too

Figure 4.1. Oron Catts and Ionat Zurr, *The Semi-living Steak*, 2000, part of the Tissue, Culture & Art Project. *Tissue Engineered Steak No.1: A Study for Disembodied Cuisine.* Prenatal sheep skeletal muscle and a degradable polyglycolide acid (PGA) polymer scaffold.

Figure 4.2. Oron Catts and Ionat Zurr, *The Semi-living Steak*, 2000, part of the Tissue, Culture & Art Project. *Tissue Engineered Steak No.1: A Study for Disembodied Cuisine.* Prenatal sheep skeletal muscle and degradable PGA polymer scaffold.

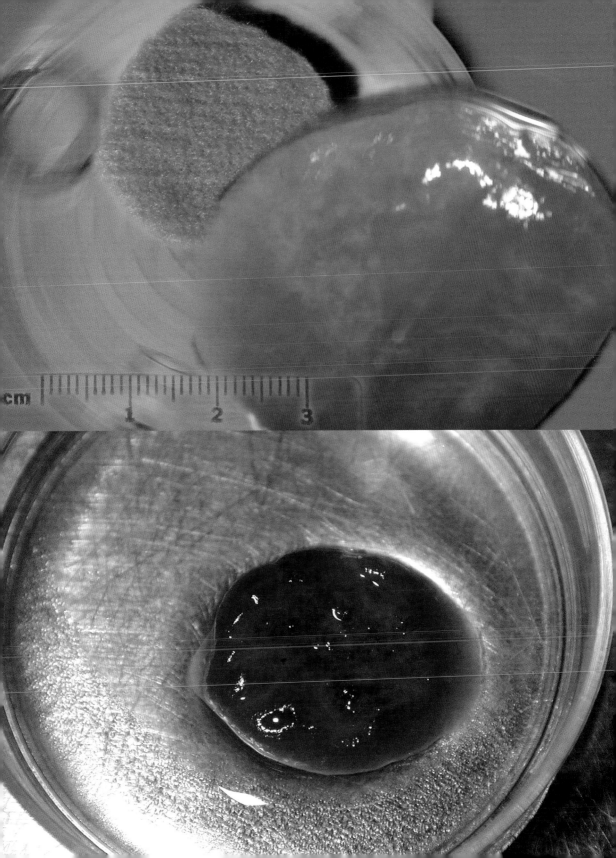

advanced for what was possible with the biotechnologies available. Yet this futuristic statement presented a vision of a world where science would remarkably revolutionize how we produce food. How much closer to lab-grown meat are we today?

In vitro cultivation of muscular fibers was achieved in the 1970s.[30] The first time a tissue culture was used exclusively for growing meat and producing food was in 2000, when artists Oron Catts and Ionat Zurr created *The Semi-living Steak* as part of the Tissue, Culture & Art Project (TC&A).[31] One-centimeter steaks were grown using skeletal muscle cells taken from an unborn sheep and then incubated in a 3D tissue culture bioreactor.[32] The cells were fed with fetal bovine serum (FBS), a necessary nutrient for tissues grown in vitro that is made from blood extracted from calf fetuses that die during the process. It is an elaborate process where the costs are high, from $600 US and upward for one liter, according to data from 2020. Animal welfare is an ethical dilemma and also comes at a price when producing FBS, considering that up to three fetuses are required to produce each liter of serum.[33]

The Semi-living Steak is more than an exploration of future food solutions. The project focuses on the ethical implications and dilemmas of using lab-grown tissue as a food source. In 2003, the project was continued further with *Disembodied Cuisine*, which grew in vitro frog steaks from cells taken from the biopsies of live animals. No frogs were killed in the process, and in fact, they were present during the art exhibition in France, where the lab-grown frog steaks were served.[34]

Ten years later, Dutch scientist Mark Post announced and unveiled the world's first cultured beef burger, often referred to as the *frankenburger*. In 2013, the cost for growing this food item was $332,000. The costs were partly due to the fifty liters of FBS needed to produce the burger. In effect, it can be estimated that somewhere between a hundred and three hundred unborn calves had to be slaughtered to make the first cultured beef burger. In 2019, Post claimed that the cost per burger had fallen to $10.[35]

If lab-grown steaks and burgers will be the solution to our environmental problems in the future, do they also taste good? According to Catts and Zurr, as lab-grown muscle does not get any exercise in their petri dishes, there are several issues with producing lab-grown meat. It is a challenge to produce the meat thick enough and with the right consistency and texture.[36] The taste might also be a bit off. As Catts describes it: "the steaks tasted like froggish jelly."[37]

To explore the culinary properties of the novel meat alternative, Mark Post organized a tasting for the first lab-grown burger in 2013. It was cooked during a live TV food show and tasted by two food writers, who felt that the main difference between the real burger and the lab-grown burger is the lack of fat in the fake burger. Otherwise, the lab-grown burger tasted similar to a conventional meat patty. The burger served was made by mixing the lab-grown meat with bread crumbs, egg powder, and some spices. The coloring of the pasty-colored stem-cell strands was improved using beetroot and saffron.[38]

The in vitro food experiments reveal many unresolved technical complications, ethical dilemmas, and even culinary challenges in making cellular agriculture part of large-scale food production in the future. The main issue revolves around the FBS that is needed to grow these cells in the lab. The paradox is that in order to grow victimless meat, multiple animals must be slaughtered to obtain each liter of serum. The use of the term *victimless* therefore sounds both hollow and ironic. These challenges are major obstacles to scaling up the industry of lab-grown meat.

A new milestone for the cultured meat industry was reached in 2020 when for the first time, a regulatory authority approved lab-grown meat for sale commercially. The chicken bites created by Eat Just, a US company, are made by combining cultured chicken meat with plant-based ingredients. This product has passed a safety review in Singapore. While the company used FBS in the approval process, it aims at replacing it with plant-based serum for the coming production line.[39]

If plant-based serum becomes a reality, it would resolve the major challenge of making the cultured meat truly victimless. In the future, we might witness the rise of the neomnivores—those who eat only cultured meat.[40]

Fishless Fish

The environmental problems of our industrial food system reach far beyond agriculture. We are straining our land-based resources, but the problem also occurs when sourcing food from the oceans. Overfishing, the extinction of marine species, and water pollution are some of the main reasons that the innovation of fake foods has also taken on the task of finding a substitute for fish and seafood. Similar to innovations in the meat market, there have been several proposals to replace fish products with plant-based and lab-grown substitutes.

One such project testing the potential of in vitro seafood is the food startup Finless Foods, which early on experimented with growing the cells of various fish.[41] The company has tested growing cultures from zebrafish, goldfish, carp, and salmon, among others. Currently, Finless Foods is working with cells from the bluefin tuna, which is commercially not suited for farming. The hope is not only to make an environmental alternative to save the bluefin tuna from extinction but also to decrease the environmental impact of transportation and reduce the costs of the product. Another advantage of replacing the fish with a lab-grown substitute might be that the resulting product is free of the mercury and microplastics that are common components in ocean-caught seafood. As discussed in chapter 3, Anthropocene Specials, there will be more plastic than fish in the world's oceans by 2050, meaning that microplastics will be unavoidable when consuming seafood. For those not yet ready to acknowledge plastic-fed wild seafood as part of their diet, lab-grown fish meat might be the only alternative.

While lab-grown meat requires nonsustainable FBS, there is evidence that fish blood could potentially be utilized as a replacement.[42] Even then, there would still be victims—the fish.

Victimless Delicacies

Cellular agriculture promises victimless animal products to reduce the moral and ethical concerns of eating meat and the environmental issues of industrial farming. Yet the edibles presented so far look like big lumps of red cells or conventional minced meat. Could the combined efforts of cellular agriculture and 3D printing technologies open up possibilities to shape the in vitro meat into something more aesthetically pleasing?

An attractive area to investigate using in vitro technologies is food that currently requires only a small part of an animal, often rendering the rest of it as waste. Such is the case with Chinese traditional shark fin soup, which is made using only one part of the shark—its fin.[43] It is a delicacy many are willing to acquire, even if illegally through shark finning (see chapter 3, Anthropocene Specials).

In China, many endangered animal and fish species are sold not only to be eaten as delicacies but also to be used in traditional Chinese medicine.

Figure 4.3. Kuang-Yi Ku, *Tiger Penis Project*, 2018.
Organs created using synthetic biology to replace
animal products used in traditional Chinese medicine.

To explore the potential future applications of cellular agriculture in China, the artist Kuang-Yi Ku created *Tiger Penis Project* (2018). In this speculative design experiment, Ku and his bioengineer collaborators investigated whether wild animal organs could be developed using genetic modification, 3D bioprinting, and tissue culture. *Tiger Penis Project* focuses on replicating the tiger penis that in traditional Chinese medicine is believed to enhance male virility. While other cultures view the healing qualities of animal parts as a myth, in China the high demand for tigers and other wild animals used for healing purposes poses a massive threat to endangered species. The artist believes that creating a new interpretation of traditional Chinese medicine using emerging biotechnologies would secure cultural traditions while ensuring no harm is done to wildlife.

If it might be possible to replicate individual animal parts using cellular agriculture, why not take a step further and create new substances that combine several mythical animal healing powers in one? Ku recognizes that both Eastern and Western cultures have such cultural myths and that it might be possible to create foods that have an impact across different cultures. On a theoretical level, Ku speculates that a cell containing the genes of various creatures believed to enhance male virility (such as tigers, oysters, and octopuses) could be used to synthesize new foods with mythical powers.[44]

Insects

If we can grow beef, fish, and even separate animal parts in the lab, what other organisms might be beneficial to culture artificially? What about insects? They are an important part of the diets of 2 billion people around the world,[45] but their populations are rapidly decreasing (see chapter 6, Bug Buffet). Can we use cellular agriculture to ensure that this dietary practice will endure? The in vitro possibilities and flaws of lab-grown insects are explored by Oron Catts, Ionat Zurr, and Robert Foster in the project *Stir Fly: Nutrient Bug 1.0* (2016). This experiment is part of the Tissue, Culture & Art Project, but instead of growing animal-based meat, the artists created a prototype of a domestic bioreactor that cultivates in vitro insect proteins from fly cells. Unlike animal cells that need to be cultured at 37 degrees Celsius, the cells of cold-blooded insects

Figure 4.4. Oron Catts, Ionat Zurr, and Robert Foster, *Stir Fly: Nutrient Bug 1.0*, 2016, part of the Tissue, Culture & Art Project. Custom-made bioreactor, insect cells, nutrient media.

can be cultured at room temperature, making it a potentially viable solution for domestic protein production outside a lab.[46]

Toward Fake Foods

Even if fake foods and artificial food substitutes are the answers to our future food scarcity and environmental issues, how do we help people shift away from their traditional diets, especially the animal protein that has shaped human society and anatomy profoundly for over 2.6 million years?[47]

And how can we trust the quality and food safety of these faux foods when there are many cases of seriously hazardous and even fatal fake foods? As mentioned above, the increasing demand for dairy products in China led to the production of nutritionally insufficient and even hazardous products. Commercially sold infant milk formula gained popularity among lower-income Chinese people,[48] which led to the decline of breastfeeding and increased use of the low-quality infant formula produced by several brands. At least thirteen babies died from malnutrition after consuming fake milk made when dairy farmers watered down cow's milk to increase their produced volume and therefore profits. After the scandal, tighter control was imposed on overseeing the nutritional value of milk products. Yet due to low profits, farmers found other ways to cheat, such as adding melamine to already watered milk to give the appearance of higher protein content. This led to another scandal in 2008 when melamine-laced dairy products in China sickened 53,000 young children and killed at least four who died from kidney failure.[49]

Fake foods have a bad reputation as both their origin and content cause suspicion. Yet to feed the world in the Anthropocene, we cannot count only on traditional food production and natural resources anymore. While overcoming the taboo of eating fake foods might be a long cultural and social transition, it is also known that "The cheapest and most reliable technique for changing a food system is to make certain foods taboo."[50] Therefore, the increasing adoption of vegetarian and vegan diets might hasten the transition to a future free from animal products. According to *The Guardian*, by 2050, there's a good chance that it will be socially unacceptable to eat meat.[51] And why should we eat meat if we can develop meat substitutes that taste better and are healthier?

Figure 4.5. *In Vitro Oysters* from *The In Vitro Meat Cookbook* (2014) by Koert van Mensvoort and Hendrik-Jan Grievink.

Figure 4.6. *Knitted Meat* by Alberto Gruarin from *The In Vitro Meat Cookbook* (2014) by Koert van Mensvoort and Hendrik-Jan Grievink.

Although using cellular agriculture to grow artificial meat is still in its early development, there are many potential uses to explore. An inspirational source for perfecting lab-grown meat cooking skills is *The In-Vitro Meat Cookbook* by Koert van Mensvoort and Hendrik-Jan Grievink.[52] This cookbook for a meatless future presents forty-five imaginary recipes that might be part of our cuisine to come. The book consists of such recipes as *Postburger* (inspired by the hamburger Mark Post made in 2013), *In Vitro Kebab*, *Meat Powder*, *Vegan Gelatin*, *Bacon on a Roll*, and even *Cruelty Free Pet Food*. While today *The In-Vitro Meat Cookbook* serves only as a thought-provoking concept that provides little help in the kitchen, in the future, it might become more useful than a cookbook containing traditional animal-meat dishes.

Looking even further into the food future, how many real foods can we replace with fakes? Could a healthy and tasty future diet contain only artificially produced food? The Japanese conceptual design studio Open Meals, Takashi Koyama, and Ryosuke Sakaki claim that the fifth step of the food revolution will be the digitization of food, or as they call it, *food singularity*.[53] Open Meals aims to create the world's first standardized digital food platform, *The Food Base*, where it will be possible to upload, distribute, and download food data for every dish. The collection of flavors, shapes, colors, and nutritional values will allow cooks to reproduce any dish in the world through the *Pixel Food Printer*. For the first test meal, Open Meals selected a traditional Japanese hot pot dish (oden) and turned it into a 3D printed artificial version (*Digital Oden*) using a flavorless, odorless gel as a base ingredient to reproduce the original meal.[54] *Pixel Food Printer* resembles an early version of the replicator from the *Star Trek* movie franchise—a molecular synthesizer device using matter-energy conversion technology to produce any food desired. Could such concepts as *Pixel Food Printer* and even the replicator become everyday kitchen utilities in the future?

Food Printers

The 3D printed foods discussed in this chapter, such as *Egg Printer* and *Tiger Penis Project*, are conceptual and speculative but might be able to be manufactured in the future. While a still relatively new technology that has developed

Figure 4.7. dentsu / Open Meals, *Digital Oden*, 2018. 3D models of ingredients for oden—the traditional Japanese hot pot dish.

Figure 4.8. dentsu / Open Meals, *Digital Oden*, 2018. 3D printed ingredients for oden.

over the past fifteen years, 3D food printing is gaining momentum both in the food industry and academic studies. Could this additive manufacturing technique be used to print all food in the future?

The possibilities are nearly endless. As long as an ingredient can be deposited as a paste or powder layer by layer through the printer's nozzle, it can potentially be printed.[55] It is also possible to manipulate hard materials by melting them during the extrusion.[56] Even if it prints, do we necessarily need it? 3D printed chocolate confections or even the NASA-funded 3D printed pizza toppings[57] are unlikely to contribute much to our future survival. Can 3D printing play a more meaningful role in producing food for us in the Anthropocene?

The aesthetic qualities of 3D printing can become useful when introducing and normalizing novel foods. Such an innovative approach is taken by Susana Soares in the project *Insects Au Gratin* (2011), which reinvents insect-based cuisine by creating unusual-looking edibles. Printed from insect flour, they do not resemble creepy bugs but can take any shape desired (see chapter 6, Bug Buffet).

3D printing can be also used to manufacture highly realistic fakes using alternative ingredients. Hanan Alkouh, in the project *Sea-Meat Seaweed*, uses molding to produce vegetarian faux meat, and the Italian company Novameat manufactures realistic steaks and pork chops through high-resolution printing.[58] Instead of unaesthetic veggie burgers, the result is elaborate three-dimensional fakes.

Additive manufacturing can make future synthetic cuisine more appealing. Designer Meydan Levy in the speculative project *Neo Fruits* (2018) uses 3D printing to create a collection of artificial fruits for a future scenario when all food has to be created artificially.[59] Each of the five different fruits is made by 3D printing translucent cellulose skins in a flat form. These skins are later filled with a mix of vitamins and minerals with distinct tastes and colors for each fruit.[60] The printed coating gives the fruit its texture and fruit-like form. The project envisions future food supplements that are functional and offer a sensory eating experience that mimics the eating of real organic fruit.

These projects show that 3D printing can serve as an essential tool for enhancing eating experiences, producing realistic fakes, and normalizing natural and synthetic novel food ingredients that might become a fundamental part of our future diet. Such fakes could taste, look better, and even be more sustainable than the original dishes.

Summary

Meatless meat. Fishless fish. And milk without cows. If consuming traditional foods of organic origin will no longer be viable or culturally accepted, we might have to replace the real foods with fake foods. In the current transition toward the artificial and fake, novel foods are being introduced to be as close to the original dish as possible. If we succeed in mimicking the original foods down to perfection, we can start using radically different resources to produce them.

Fake foods could be beneficial to food production in times of ecological crises. However, urban legends from around the world make this a controversial topic. While cheese produced from umbrella handles and eggs made from synthetic resin have turned out to be hoaxes, others, such as the dangerous faux milk from China, had serious health consequences and resulted in fatalities. However, as the Chinese wine counterfeiting industry shows, taste can be culturally acquired, making the fake more popular than the original.

Imposter foods could replace animal-based products that have such an enormous negative impact on the environment. Innovative alternatives to meat range from bioengineered protein that resembles heme found in meat to natural alternatives such as bacon-flavored algae. There is also the potential for growing a variety of cultured meat products from the cells of mammals, birds, frogs, fish, and even insects. Cellular agriculture in combination with 3D printing could even eliminate the illegal market for endangered animals used in traditional Chinese medicine. Eventually, high-resolution 3D printing might allow us to fake any shape, taste, and nutritional value of desired food.

While the cellular agriculture industry is yet to resolve the main technological and ethical issues of lab-grown meat, recent developments point to progress. With the first lab-grown meat approved to be sold commercially in Singapore, it is possible that fake food hoaxes (such as the in vitro celebrity meat sausages by Bitelabs) will one day be part of the food market.[61] If we are willing to play with the possibility of eating lab-grown human meat, why not look into other culinary possibilities the human body offers?

Human Deli

5   Human Deli

In an age of catastrophes and unforeseens, everything that can be eaten must be investigated as a potential food source for humans, even the human. The living urgency of the Anthropocene means that the current menu is about to expire and soon will become too resource-intensive or perhaps even outright impossible to produce in the future. Looking for alternative resources to replace our favorite staple foods is a necessity.

Adding to the negative environmental impact of current industrial food production are the high costs for the global distribution and transportation of food. These costs add massively to the bill for future food crises. Food transportation is an energy and resource-intensive part of the global food system. Today the world is highly dependent on the global trade of goods. In a world with an unequal distribution of wealth, many would go hungry without trade in staples that are basic for nutrition, such as grain, meat, and dairy. Adding to this are the massive numbers of trendy foods that are shipped across the globe to tickle the taste buds of wealthy customers but often leave behind a negative environmental impact. What if we were no longer able to import exotic foods from overseas—or worse, any food at all?

To replace our appetite for globally imported exotic foods, we must look for local alternatives that provide new delicious flavors and even some additional nutrients. Our understanding of self-sufficiency and sustainability must reach far beyond the ecoconsumer fashion of buying discounted leftovers from high-end restaurants or dumpster diving. How local can we go in our food production? What if we become a food source ourselves?

The human body carries a lot of potentially edible material. Could the trend toward locally sourced organic foods create a perfect storm for edibles produced using human material? This chapter explores creative ways of using human by-products—such as breast milk, saliva, various bacteria, seminal fluid, urine, feces, and even human flesh—in food production. All the foods presented in this chapter are a result of hands-on experiments. Several of the meals have been offered and consumed as edible experiences with audiences.

Human Bacteria

The exploration of the human body's potential role in food making starts with looking at the part of the body that is invisible to the naked eye—bacteria. While bacteria raise associations with disease and uncleanness to many, it would be impossible for humans to live in a bacteria-free environment. The human body is estimated to contain equal numbers of bacterial cells and human cells.[1]

Also, our food is nowhere close to being bacteria-free. Bacteria are essential to nearly all commercially produced fermented foods. They are used in starter cultures to enhance preservation, to improve nutritional value, and to achieve the desired flavor, aroma, texture, and appearance.[2] Can this knowledge help us to utilize the rich human microbiome in food production to create new unique foods and flavors from our bodies?

How to use human bacteria in food making is explored by biologist Christina Agapakis and olfactory researcher Sissel Tolaas in the project *Selfmade* (2013). *Selfmade* resulted in a collection of cheeses made from human bacterial starter cultures. First, Agapakis and Tolaas collected bacteria samples from human hands, feet, noses, and armpits. The swabs were inoculated into pasteurized milk and incubated at 37°C, the same temperature as human body heat. The result was a series of unique-smelling fresh cheeses that "are scientific as well as artistic objects, challenging us to rethink our relationship with our bacteria and with our biotechnology."[3] Another line of cheeses was developed using the bacteria cultures of celebrities. Agapakis and Tolaas made cheeses from bacterias in artist Olafur Eliasson's tears and author Michael Pollan's belly button.[4]

The authors of the project expose how different microbial populations are part of the cheese-making process. Cheese quality and characteristics depend on many factors, such as quality of milk, culturing process, and aging. Agapakis and Tolaas speculate that many unique cheese flavors might have been

Figure 5.1. Christina Agapakis and Sissel Tolaas, *Selfmade*, 2013. Human bacteria sample cultured in a petri dish.

Figure 5.2. Christina Agapakis and Sissel Tolaas, *Selfmade*, 2013. Cheese made from microbiologist Ben Wolfe's toe microbes.

achieved due to the presence of human bacteria, which is hard to avoid in the cheese-making process. *Selfmade* highlights the similarities between the human bacteria forming the smell of human armpits or feet and the sources of the scent of some of the stinkiest cheeses.[5]

While the cheeses were not intended or offered for consumption, the project's authors claim that they may be safe to eat.[6] We can only speculate on how the various starter cultures of human bacteria influenced the resulting taste. Added microbial cultures and secondary microorganisms play an important role also in the development of cheese flavors during the initial steps and ripening.[7] The resulting product's characteristics depend on the dynamics of all the involved microbial populations. All bacteria—both desirable and undesirable, yeasts and molds—compete during the process.[8]

The same would happen with the microbiota of *Selfmade* cheeses that have been composed using unique starter cultures from human bacteria. What are the different taste and texture characteristics of cheese made from bacteria on feet and cheese made from bacteria from armpits? And how different would nose-bacteria-based cheeses taste when made from starter cultures collected from different people? The characteristics of cheeses and any other fermented foods to an extent are influenced by the presence of human bacteria. By utilizing human bacteria, individuals can produce foods unique to their body microbiome.

Vaginal Food

Bacteria live all over and inside our bodies. The majority are on our skin and in the mouth, the gut, and the vagina.[9] A healthy female's vagina contains hundreds of different types of bacteria and organisms. *Lactobacillus* is a dominant bacteria, which is also one of the two most common probiotic species used for yogurt starter cultures. The *Lactobacillus* found in both yogurt and the vagina chew up milk proteins by fermenting glucose into lactose and produce lactic acid as a by-product.[10] Due to the unique microbial properties of the vagina, numerous artists and researchers have shown an interest in vaginal discharge as an ingredient in food making. Experiments with vaginal bacteria have also been conducted by members of the general public.

Cecilia Westbrook took the challenge and experimented with her vaginal secretions to create yogurt.[11] According to microbiologist and writer Rosanne

Yente Hertzberger, who reflected on this experiment, it is difficult to examine how exactly the *Lactobacillus* from Westbrook's vagina interacted with the milk as there must have been many other bacteria present in the process, including other vaginal bacteria.[12]

So what does food produced by human vaginal bacteria taste like? According to Westbrook, the taste of the first batch could be compared to Indian yogurt, with a sour, tangy flavor, while the second batch tasted like slightly spoiled milk.[13] Although her yogurt is probiotic to a certain degree, the fermentation process is sure to promote various other bacteria living in the vagina. For example, some vaginal bacteria could be pathogenic and pose a serious health risk if consumed in large numbers. If *Lactobacillus* is so useful in food fermentation, what other applications could vaginal secretions be used for?

Artist Toi Sennhauser used her vaginal yeast for a yeast starter culture to create bread in her project *Mama's Natural Breakfast* (2005). The unique homemade bread was on public display during the Seattle Erotic Art Festival, where it was served with butter and honey on the side. The artist sees *Mama's Natural Breakfast* as a social experiment that makes the audience question if it is really improper to ingest the material in question. Also on display was a table tag informing viewers that "eating this bread has been forbidden by the Seattle Public Health Department."[14] Meanwhile, the artist claims that the bread is food safe as she had tested negative for sexually transmitted diseases.

Bread making has also been explored by blogger Zoe Stavri, who made a sourdough bread starter using vaginal discharge collected during a yeast infection known as *Candidiasis*. The infection changes the balance of bacteria and yeast in the vagina, allowing the yeast cells to multiply more rapidly.[15]

Toi Sennhauser has also experimented with using vaginal discharge in beer brewing. In 2005, the artist created the *Original Pussy Beer*. The project was inspired by ancient female brewers who lived 7,000 to 4,000 BCE in the kingdom of Sumeria in Mesopotamia. While the beer was made through a conventional beer brewing process using grains, the artist combined the brewer's yeast with vaginal yeast.[16] The taste was described as fruity, light, and sweet.[17] Just like with cheeses with human microbiota present, beer should form a unique taste due to the various bacteria used in the brewing process. While it is difficult to evaluate how much the human bacteria affected the beer, there is one human body fluid that we all have tasted, and that is breast milk.

Breast Milk

While consuming most human byproducts is taboo and more likely to evoke disgust than demand, milk is considered a natural fluid all mammals, including humans, produce to feed their young. Several thousand years ago, humans started to evolve the ability to digest milk as adults, allowing animal milk to become an everyday staple.[18] As milk contains several essential nutrients, it has become an important part of the modern diet. At the same time, the consumption of industrial dairy products is increasingly criticized due to its environmental impact and ethical concerns for animal welfare. A growing number of consumers are switching to plant-based milk substitutes.[19] Most milk alternatives sold commercially are milk-like liquids produced from various plants such as almonds, oats, rice, soy, and coconut. A common issue with these products is their lower nutritional value compared to bovine milk. According to a recent scientific study, while soy milk, almond milk, and cow's milk have somewhat similar protein levels, the sugar levels in almond, coconut, and rice milk are considerably higher. The plant-based milk substitutes also vary in calorie, vitamin, and mineral levels, but most important, all of them lack the essential amino acids existing in cow's milk.[20]

Are these plant-based milk alternatives equal to traditional products from real dairy? Or are they just as nutritionally incomplete as the fake watered-down dairy products fabricated by Chinese dairy farmers in the 2008 scandal (see chapter 4, Fake Foodies)? Research shows that if consumed regularly, these plant-based products can have a negative nutritional impact and can even lead to micronutrient malnutrition if consumers are not aware of the essential differences between the composition of bovine and plant-based milk.[21] The booming consumption of plant-based milk substitutes as a replacement for animal milk is another example of orthorexia. Not everything that ticks the box for being sustainable and healthy is necessarily what you need. So how do we solve our dairy problem if plant-based alternatives do not provide the same benefits as cow's milk? Can we produce nutritionally complete milk without having a destructive environmental impact while addressing the ethical dilemmas about animal welfare that arise with dairy farming? It is impossible to produce milk with consent from the cows. It is, however, possible to obtain human breast

Figure 5.3. Miriam Simun, *Lady Cheese Shop*, 2011. Making human breast-milk cheese. Videostill.

Figure 5.4. Miriam Simun, *Lady Cheese Shop*, 2011. Human breast-milk cheese samples.

milk with consent. All mammal milk has the same nutritional composition. Only the proportions differ. Human milk is characterized by high water content and low energy density, making it more similar to zebra's milk than cow's milk.[22] Yet could we use human breast milk to produce our favorite dairy products without the need for domestic farm animals? With this question in mind, several artists have explored the use of human breast milk in gastronomy.

Artist Miriam Simun launched the *Lady Cheese Shop* (2011), where she invited audiences to taste various breast-milk cheeses. The main emphasis of the project was to produce cheeses from locally sourced food. Two kinds of cheese were made using donated breast milk from two women in New York, and the third was made with breast milk ordered online from another city in the United States.[23]

Similar to other projects discussed throughout the chapter, human breast milk is not the only ingredient in the final product. According to the artist, a pure human cheese is not biochemically possible as it has too low amounts of the protein casein that is needed for curdling the milk. The human breast milk was therefore mixed with goat milk. Similar to vaginal secretions used in beer brewing and bread making, using human milk presented a possibility that it would also contain unhealthy bacteria or viruses from the donor. For the *Lady Cheese Shop* event, all donated milk had to be tested for pathogens for the safety of consumers before audiences were invited to participate in cheese-tasting sessions.[24] To avoid common disgust and apprehension against human food, the cheeses were presented in palatable and appetizing ways. This contributed to making their consumption socially and culturally acceptable. Still, the eating evoked different reactions in different audience members, from laudation to regurgitation.[25]

Human milk can also be used to create such dairy classics as ice cream. As part of the *Art Meat Flesh*, Cathrine Kramer from The Center for Genomic Gastronomy prepared human breast milk ice cream. Kramer was interested in developing an ice cream that could be potentially accepted by the vegetarian community. Using human breast milk allowed her to create a dairy product with milk obtained with consent.[26]

Human breast milk products have also reached outside the arts community and been sold commercially. In 2011, the Lickators, a London-based ice cream manufacturing company, received 850 milliliters of breast milk donated by a breastfeeding mother. This allowed the company to make the first fifty servings of Baby Gaga ice cream.[27] While the ice cream was initially confiscated to undergo a food safety test, the health protection agency eventually declared that it was safe for human consumption.[28]

In a future scenario, when farm animals are no longer seen as an ethically sound source for milk, a radical approach would be to explore the potential for using human breast milk instead. If human breast milk does not pose any health hazards, does it mean these products will soon be sold at conventional grocery stores? While the Lickators succeeded in launching a commercially available product, many ethical and practical issues arise in obtaining human breast milk. In what quantities can a breastfeeding mother produce milk? And is it ethical to use a mother's milk to create cheeses and ice creams when it is initially intended to feed a newborn?

In some cases, a breastfeeding mother can produce up to two liters of milk a day.[29] In comparison, a dairy cow, on average, produces twenty-eight liters of milk per day during the lactation period. A high-yielding cow during the peak lactation can make as much as sixty liters.[30] This makes humans very inefficient mammals when it comes to dairy production. In addition, the costs of nourishing a lactating human are higher than feeding a cow with grass.

The increased milk production in cows has not developed by itself. It has been achieved through genetic modification using the growth hormone recombinant bovine somatotropin (rBST). And the alteration of the cow's genome does not stop at the amount of milk produced. Scientists have also modified the milk's nutritional composition, in some cases to make it similar to human breast milk[31] and in others to produce milk that can be safely consumed by lactose-intolerant people.[32]

If we have succeeded in genetically modifying the milk produced by a dairy cow, could the same be done with human breast milk? If we were to use humans for milk production, manipulating the human genome for an increased amount of human breast milk is a speculative possibility. And while using female breast milk for adults competes with feeding newborns, human males could be enhanced for food production purposes instead. All males have rudimentary mammae, the glandular organ necessary for secreting milk. In effect, every man can produce milk, given the right conditions. A study investigating human male lactation summarizes known cases caused either by environmental factors, induced hormonal changes, diseases, or psychological stimuli connected to feeding babies.[33] While seen as extremely rare, the secreted milk quality by males is comparable to that of the lactating females.[34]

While care for the young can induce lactation even in males, it is doubtful that humans will start producing milk due to their concerns for feeding an increasingly hungry world population in the Anthropocene. Therefore, human milk is more likely to remain constrained to breastfeeding the young and occasionally creating ice cream rather than to exploding into a large-scale dairy industry. Still, our bodies contain other nutritionally valuable and harvestable fluids.

Semen Cuisine

As a human by-product, semen can be considered a sort of a superfood due to its rich composition. It contains over two hundred different proteins[35] and other nutrients, such as magnesium, potassium, sodium, and zinc.[36] Is this bodily liquid just as valuable in culinary qualities as it is in chemical composition? Semen-based recipes are presented in the book *Natural Harvest: A Collection of Semen-Based Recipes* (2008) by Paul "Fotie" Photenhauer. The author describes seminal fluid as an inexpensive and commonly available resource that has a complex and dynamic taste, wonderful texture, and excellent cooking properties.[37] The cookbook format serves as a playful yet powerful instrument for introducing the edibility of this ingredient. As the book offers real and tested recipes, it invites readers to imagine the human body as a food source.

Do its rich nutritional value and culinary qualities make semen a valuable edible resource for our survival in the Anthropocene? On a theoretical level, semen is a local and organic resource that can be harvested rather easily. Yet is it possible to obtain enough of it to contribute to food security? There is even an urban legend that relatively more women than men survived the World War II siege of Leningrad by consuming semen.

Will seminal fluids become a nutritious emergency staple for the food crises of the future? The average volume per ejaculation is 3.4 milliliters.[38] The protein concentration in this bodily fluid is around 5,040 milligrams per 100 milliliters,[39] which means that one ejaculation equals only 171 milligrams of protein. To put this in perspective, to obtain protein equal to the amount in an egg (13 grams), one would need to obtain seventy-six portions or ejaculates of semen. A decreased nutritional intake during food scarcities, such as during the war and famine, would likely critically reduce the quality and quantity of what can be harvested. It is, therefore, unlikely that semen will become an emergency staple food, not because of the cultural taboo around consuming seminal fluids outside a sexual context but because of the very low volumes obtainable.

Even though seminal fluid cannot be sourced in quantities large enough to be a valuable food source, it can still be used as an exotic but locally sourceable ingredient to, for example, enrich the flavor of foods.

Golden Drinks

The largest amount of fluid that the human body discharges on a daily basis is urine. And as with the consumption of seminal fluids, urophagia or the practice of consuming urine is perceived as taboo by most, yet it has a long cultural history. Urine, also referred to as *gold of the blood* and *elixir of long life*, was consumed by ancient cultures from Egyptians to Romans mainly for various therapeutic purposes.[40] Some still practice urophagia today, and in popular media, it is considered a last-resort source of liquid in case of emergency. Yet many survival guides, including the *US Army Field Manual* (2002)[41], advise against it. According to the manual, urine contains 2 percent salt, which would amplify dehydration. However, due to urine's high water content and rich chemical composition, it can be recycled and processed using advanced technologies to reclaim various valuable byproducts, including clean drinking water.

Although we on earth are struggling to overcome the taboo of drinking reclaimed water from human waste, astronauts have been consuming distilled urine in space since 2009.[42] Each person discharges up to two liters of urine a day. With 91 percent to 96 percent of urine being water,[43] squandering it is simply a waste of one of the most precious natural resources in space—water. The closed-loop system onboard the International Space Station recycles up to 93 percent of all water sources such as sweat, urine, exhaled water vapor, and other wastewater.[44] The gut-wrenching question many ask is, What does the end product taste like? The answer is "aside from a slight tang of iodine, it tasted like, well, water."[45] It is also most likely much purer than any tap water drunk on earth.

Urine also contains various organic and inorganic materials, such as calcium, magnesium, and glucose, among others.[46] Examining the composition of urine is a common way to study the use of pharmaceutical products. It is even possible to reclaim some of them. In early trial cases of using penicillin, doctors recovered unabsorbed medicine from the patient's urine for reuse.[47] It is not the only way to utilize urine. It can even be used to produce an exquisite and expensive liquor.

The UK-based designer and researcher James Gilpin was inspired by the fact that people with diabetes have high sugar levels in their urine. Large amounts of reusable carbohydrates are, therefore, simply going down the drain to waste every day. As both himself and family members have the condition, Gilpin went about trying to utilize this rarely considered resource and created a unique product out of it—the *Gilpin Family Whisky* (2010)—a single malt whiskey developed by processing urine obtained from older diabetic patients.[48] Specifically, Gilpin extracted the excess sugar molecules from the urine concentrate and then added them to the mash to speed up fermentation. While the project's primary focus is on the health conditions of patients with diabetes, the designer speculates on how to utilize the water purification systems to harvest the biological resources humans produce in abundance.

Eating human bodily materials can be perceived to carry health risks. However, by using advanced techniques and processing methods, the concerns for food safety can be, at least in theory, removed. Then the discussion can move directly to the culinary qualities and nutritional value of the ingredient in question. *Gilpin Family Whisky* challenges the taboo of consuming urine by breaking its composition down to basic chemical compounds and finding a practical use for it.

Fecal Meals

While recycling urine into clean drinking water is an increasingly accepted practice that can improve water security worldwide, humans also discharge another reusable waste in large amounts daily—that is, shit.

Everyone eats. And everyone shits. On average, humans excrete 128 grams of feces per person per day.[49] Fresh feces contain about 75 percent water; the remaining is up to 93 percent organic solids.[50] What we consider to be waste has plenty of useful undigested material to reclaim. Around 50 percent of the original energy in the ingested food is still left in fecal matter.[51] These organic solids consist of bacterial biomass, protein, carbohydrates, and fat, making excrement a potentially valuable human by-product from which to extract both clean drinking water and reusable and reedible food.[52]

What might seem like the worst dystopian food future to many is a real scenario currently investigated by scientists and researchers for one of the most extreme environments we know—outer space. The National Aeronautics and Space Administration (NASA) has funded research to test the concept of turning astronauts' waste into food.[53]

An experimental approach to turning poop into food is to let the waste undergo anaerobic digestion—treating waste with microbes to recycle the waste back into an edible mass. According to the researchers, the current outcome is a microbial goo resembling Marmite or Vegemite.[54]

But why consider a closed-loop efficiency only in space when our own Spaceship Earth is increasingly struggling to sustain our dietary demands? In theory, the excrement of 8 billion people could be recycled into up to 69,000 tonnes of protein and 59,000 tonnes of carbohydrates every day. So are we wrong to treat human waste as waste in a world that is increasingly concerned about being too wasteful? Could poop become your next favorite steak?

In 2011, a YouTube video showing how to synthesize meat from sewage circulated on the internet. The story was about the Japanese scientist Mitsuyuki Ikeda, who created meat from the sewage in Tokyo.[55] In the video, the scientist claimed he was asked by the Tokyo sewerage bureau if he could devise a way of using up the city's excessive sewage mud. Mitsuyuki claimed to have extracted proteins from the waste while adjusting the desired meat pigment with food coloring and the flavor with soy.[56]

This controversial news story turned out to be a food hoax. The video that was cited by several online news platforms was soon taken down as no solid evidence was found about the existence of the Japanese scientist. No sewage steak was ever made. Yet is theoretical speculation about using human feces to solve the global food crisis really absurd? Our near catastrophic future might prove that our waste is much more than worthless. Rebranding feces as biosolids might help to acknowledge stool as a valuable product much needed by the farming industry in particular. As an example, in eighteenth-century Japan, before the production of fertilizer, biosolids were considered not waste but expensive, sought after, and fought for resources necessary to grow rice.[57]

Human substances are not just waste. In addition to naturally discharged human by-products such as breast milk, feces, and urine, there are modern methods to extract excess materials (such as fat) from the human body that might be turned into useful proteins.

Figure 5.5. James Gilpin, *Gilpin Family Whisky*, 2010. Whisky created using the sugar contained in the urine of older diabetic patients.

Human Waste as a Protein Source

While the global population is requiring increasingly more food to sustain itself, according to a World Health Organization (WHO) report, in 2016, more than 1.9 billion adults were overweight, and of these, 650 million were obese.[58] Alongside an increasingly overweight population, a lot of human material is being removed through plastic surgery every day. In 2019, the fat-removal procedure known as liposuction was the second most popular cosmetic surgical procedure performed worldwide.[59] Every year, close to 2 million liposuction procedures are performed to remove fat from various body parts.[60] During a single surgery, it is safe to extract up to five liters of fat from a healthy person weighing 100 kilograms.[61]

Artists can use the transgressive power of art to investigate the ethicality of cannibalism as a means of obtaining food. Several art projects have explored criteria that could make this last-resort food source socially acceptable.

In the work series *Assimilation* (1997–2016), Croatian artist Zoran Todorović visited plastic surgery clinics to collect human waste material as an edible resource.[62] The artist obtained the waste that is usually thrown away and instead turned it into a dish and served it to the public. The first collected material was from a face-lift surgery where mostly small pieces of skin and fat tissue were cut away. The collected material could, therefore, not be used in a meat-heavy meal. Instead, the artist worked together with professional chefs who suggested making an aspic dish out of the obtained material.

Todorović has since prepared the dish in many countries and received varying kinds of feedback from the public. In Croatia, Serbia, and Slovenia, the artist observed that many audience members wanted to taste the meal. In Germany, the public was mainly concerned about the legalities of cannibalism, not the taste. In Great Britain, health was the main issue as people were prohibited from eating the food because a sanitary certificate had not been issued.[63] This transgressive artwork challenges some of the fundamental ethical beliefs that modern society has agreed on. By offering the audience a taste of a victimless human meat dish, the artist asks us to reevaluate the taboo in a new context and possibly transgress it.

Italian artist Marco Evaristti problematizes a modern consumer culture in which we first ingest too much and then simply buy our way to slimness through liposuction. In the project *Polpette al Grasso di Marco* (2006), the artist prepared meatballs from a can filled with his own body fat obtained through plastic surgery. Evaristti uses the shock value of modifying his own body for the purpose of food to address questions of a global scale. The artist, however, did not serve the dish to the audiences but consumed the spaghetti with the meatballs by himself, performing an act of autocannibalism, the eating of oneself.[64]

Autocannibalism

There are several incidents of autocannibalism found in medical literature, more often related to drug abuse or crime than a need for nutriments. Serbian artist Marko Marković takes the investigation of autocannibalism to a new level in his performance series *Selfeater* (2009). The inspiration for *Selfeater* originated both from Marković's research on cannibalism in spiritual, necromancy rituals where deceased ancestors are eaten to maintain the spirit and from cannibalism associated with the lack of food for survival.

In the performance *Selfeater/Hunger*, a medical nurse cuts and removes a piece of flesh from Marković's left arm. The detached material is then served on a plate and consumed by the artist himself. Marković also sees his project in an ecological context where many people are willing to sacrifice their freedom for the benefit of the environment, thereby acquiring a martyr-like status. As in autocannibalism, this is bound to end catastrophically.

In a second performative work of the *Selfeater* series, *Thirst*, the audience is first served cotton candy to establish an intense sweetness in their mouths. Then a nurse inserts a hypodermic needle in the artist's arm. The needle is connected to a long tube that allows the artist to drink his own blood directly from his own arm.[65] Marković then drinks the blood while trying to communicate with the audience by maintaining a direct and stern look at them. The artist sees his work as a necessary revolt intended to change society and improve the values distorted by everyday consumerism.

Eating human flesh is one of the most dystopian visions of the future. The science-fiction classic *Soylent Green*[66] portrays a society that struggles to feed a population that has grown beyond the limits of sustainability. In *Soylent Green*, the algae-based miracle food keeping the famine at bay turns out to be made from reclaimed human corpses. *Soylent Green* displays a bleak image, pitting our future survival against the most profound ethical norms of modern society. If there are ways to source human meat without resorting to necromantic cannibalism, can it actually become a valuable additional protein source?

The Nutritional Value of Human Meat

Can human meat be more than an occasional provocation to make us talk about future food scarcity? In 2019, Swedish professor Magnus Söderlund caused a worldwide scandal by proposing eating human flesh as a possible strategy to survive global warming and climate change.[67] Even if we were to eat human flesh, could it compete with industrially grown farm animals as a renewable and sustainable source of protein?

It is estimated that a fully grown human male (eighteen years old or over) with an average weight of 66 kilograms can yield 55 kilograms of edible mass. The total nutrition value of fat and proteins would provide 125,822 calories.[68] This is in itself not particularly high when compared to other mammals similar to humans in size.[69] Also as Steven Vogel, a biologist at Duke University, notes, humans grow much more slowly than chickens and other herbivorous domesticated mammals.[70] A modern broiler chicken is already fully grown in five to seven weeks.[71] Humans also need more expensive and energy-intensive food than corn. This makes growing a full human for protein an energy- and time-intensive process that yields little to nothing from a renewable and sustainable perspective. Cannibalism also holds the risk of transmitting pathogenic diseases.

The Health Risks of a Conspecific Diet

Conspecific consumption of humans might increase the chance of survival in cases of extreme scenarios and short-term disasters when no other foods are available. Apart from the pathogenic risk of eating humans, there is evidence that a conspecific diet—that is, eating one's own species—can be beneficial for growth rates and survival.[72] However, it can come at the price of contracting a deadly disease. Consuming the human brain can lead to rare but fatal degenerative prion disorders affecting the brain, such as Creutzfeldt-Jakob disease and Kuru.[73] The dangerous consumption of human brain tissue by early humans might have also led to the genetic evolution of a prion protein gene that still offers modern humans some protection against prion diseases.[74] Another risk is that blood-borne infections and tapeworms are thought to be transmitted through cannibalism.[75]

Figure 5.6. Marko Marković, *Selfeater/Hunger*, from the performance series *Selfeater*, 2009. The artist eating a piece of flesh cut from his left arm.

If consuming humans as meat can lead to serious illnesses, would this change if we could grow meat from human cells in the lab? And could cultured human meat compete with genetically improved super animals like the broiler chicken?

Lab-Grown Human Meat

Growing human meat in the lab might eliminate the risk of transmittable diseases that persists through cannibalism.[76] As with any cultured meat, it is possible to select cells that show no presence of the feared diseases.[77]

Eating human meat is taboo-ridden, and the ethical questions and dilemmas that it poses are tougher than those raised about eating meat of animal origin (as discussed in chapter 4, Fake Foodies, and chapter 6, Bug Buffet). What if there were moral and socially acceptable methods to harvest human meat? Apart from the currently high costs, growing meat cells from humans in the lab is a scientifically established, relatively uncomplicated process that could take away much of the taboo related to eating human products. Cells in a petri dish are not considered to be living entities in themselves but rather semiliving fragments of complex bodies.[78]

As discussed in chapter 4, Fake Foodies, cellular agriculture comes with a future potential to grow beef, fish, and even insect meat without the need to harm and kill the animals. While this could potentially resolve ethical issues of slaughtering animals, we would still be eating meat grown from animal cells. Could we entirely move away from animal-based products and culture our own meat in a petri dish instead? Is it possible to source higher amounts of human meat through the use of emerging biotechnologies?

While traditional forms of cannibalism require obtaining each gram of flesh from the human body, lab-grown meat would not. The concept of lab-grown human meat gives a new spin on cannibalism—offering victimless human meat. Could this innovative approach to cannibalism allow us to become truly self-sufficient and self-sustainable? Would you grow human meat for consumption—even from your own cells?

Artist Theresa Schubert in her project *mEat me* (2020) explores such a scenario by growing muscle cells taken from a biopsy of her own thigh. Schubert

Figure 5.7. Theresa Schubert, *mEat me*,
2020. Performance.

Figure 5.8. Theresa Schubert, *mEat me*,
2020. A biopsy on the artist's leg.

is concerned with the environmental impact of industrial farming, while the project also explores and critiques the clean meat promises of the in vitro meat industry.[79] In her search for animal-free alternatives, such as plant-based nutrients, Schubert identified a lack of solutions for replacing the fetal bovine serum that is currently needed to grow in vitro meat. Plant serum techniques are still too early in their development. As an alternative solution, Schubert made a nutrient serum from her own blood. While the process was successful and no fetal bovine serum was used, the cells showed a slower growth rate.

The artworks conducted at the intersection of art and science can add an important technical layer to the debate about possible solutions to food production in the future. Besides questioning the ethicality of lab-grown human meat, *mEat me* succeeds in uncovering technical complications that are usually hidden in the labs of scientists.

With the potential of replacing fetal bovine serum with plant-based or other victimless alternatives in the future, lab-grown human meat could become a feasible way to source proteins. And the word *cannibalism* could mean something entirely different. Speculative ideas such as Bitelabs' lab-grown meat from celebrity cells would no longer be a laughing matter. But the next ethical question on the agenda is, Whose cells should we eat?[80]

Becoming Food for Others

Having reconciled ourselves to a taste for new dishes based on human by-products—that is, by humans for humans—we must also question using the human body as a food source for other species. What is our role in the future food chain? Humans, both alive and deceased, already are food for nonhuman life forms. Alive, our bodies provide necessary nutrients and great meals for a wide range of species, including ticks, mosquitos, bed bugs, mites, flies, and lice. If exposed in the wild and open, a human corpse provides excellent food for anything from vultures to wolves, bacteria, fungi, and flies. What if we placed ourselves in a greater ecological, sustainable, and culinary perspective and made ourselves taste even better for other species and therefore more valuable?

The Center for Genomic Gastronomy asks this question. Its project *How to Flavour Our Tears* (2016) is performed as an experimental restaurant offering nutritious human-based meals for other species.[81] The project inverts the food

Figure 5.9. The Center for Genomic Gastronomy, *To Flavour Our Tears*, 2016. *Eyephones* simulating the feeling of moths drinking human tears.

chain as we know it, placing humans as a desirable nutriment lower in the chain. The speculative design project proposes new tools, recipes, and rituals for humans to harvest themselves, turning the human body into a variety of new delights through the art of Autogastronomy—the art of flavoring oneself well. Further, they propose the new study of Altergastronomy—the study of human body parts as ingredients for other organisms.[82]

One of the Center's concepts is the *Moth Bar*, which creates an environment where human visitors can donate tears to feed thirsty moths. According to the project's authors, moths that usually feed on nectar and other sugary substances can experience a deficiency of salts vital to their survival. Some insects, such as the moth *Lobocraspis griseifusa* from Southern Asia, obtain the necessary salts by drinking tears from the eyes of the water buffalo.[83] *Moth Bar* consists of several tools that help audience members to flavor, induce, and capture their own tears for the consumption of moths. For example, *Eyephones* allow you to "simulate the feeling of moths feeding around your eyelids" through a set of buzzers placed on your eyelids. At the *Fat Flavouring Lab*, visitors can improve their own taste qualities by flavoring not only their tears but also other bodily materials such as fat, skin, blood, sweat, and even urine.[84]

So how would our everyday life be like if we were no longer at the top of the food chain? And what if more animals than just the occasional moth would drink our tears? Humans' place in the food chain is boldly challenged in The Center for Genomic Gastronomy's *AlterGastronomy VR* room, where visitors can experience humans being consumed by nonhuman species such as a wolf, a vulture, a maggot, and even a microorganism. The *AlterGastronomy VR* lets participants hunt and consume humans from the perspective of the chosen nonhuman species.[85]

With these speculative prototypes, The Center for Genomic Gastronomy challenges us to consider the gastronomic needs of the nonhuman species that consume us and will continue to include our bodies in their diet in the future. While we are concerned about ensuring our own survival and having positive culinary experiences with easy preparation methods, we can also enhance our own flavor to taste better for the nonhumans who utilize our bodies as their food.

Summary

The human body is full of edible resources varying in flavors, textures, aromas, cooking qualities, and nutritional values. Seeing the consumption of human bodily matters as a socially acceptable activity would offer a wide variety of unique yet locally available and easily sourced edible ingredients.

Some of the discussed organic human materials—such as semen, breast milk, and bacteria—can spice up our diet and even contribute to important processes in food production. Human bacteria can be beneficial in the fermentation of cheeses to alter their flavor, smell, and texture. Probiotic bacteria in a female's vagina have been used in yeast starter cultures to produce bread, yogurt, and beer. Human breast milk is proposed as nutritionally complete milk without destructive environmental costs and the ethical dilemmas of farming animals. Human breast milk ice cream has even been sold commercially. And while semen cannot be obtained in large enough amounts to contribute to daily nutrition, it has unique culinary properties.

Other bodily matters, such as human feces, can play a more significant role as a nutritional food supplement, substantially contributing to future food security. All human urine could be recycled into clean drinking water and valuable resources such as sugar and unabsorbed medicine. Human feces holds both precious water and undigested material, including bacterial biomass and proteins.

Using modern methods, human excess materials can be turned into a valuable protein source. Human waste, removed through plastic surgeries, including fat from liposuction procedures, can be recovered and used as food. Even more, growing human meat in the lab could give an innovative spin to the benefits of a conspecific diet by providing victimless protein without the risk of transmittable diseases. If our future diet is to be nutritious yet full of new exotic culinary experiences, human byproducts can be valuable.

Humans are not only eating. They are also eaten. By twisting our position in the food chain, we can make ourselves into a valuable food source for others and can feed other organisms, such as insects. The next chapter discusses how—and if—insects can be the next coming planetary buffet and most important food source for humans in the future.

Bug Buffet

6 Bug Buffet

Bugs are gaining attraction as the new green protein source. The race for eating insects started in 2013 when the United Nations published a report that emphasized the need to investigate and use them as a sustainable food source. That report has been downloaded more than 7 million times.[1] It makes sense. Bugs are small, lean, and mean, the latter in the sense of being competitive with current methods of meat production. Growing meat through livestock comes at great environmental costs. It is not sustainable and even less ecological. What if we could scale down the need for resources while scaling up our access to meat-based protein? That might happen if we change from livestock to microlivestock—that is, bugs. But change is not always easy, especially not if your food is creepy.

Earthworm jerky? Cricket tacos? Insect burger? Mixed bug kebab? These foods will sound unfamiliar to many and disgusting to more. Will insects really replace meat in the future—and are these our soon-to-be favorite high-protein staples? That might be just as much a question of cultural acceptance as a technological adaptation to bug farming. And there might be no other choices in the Anthropocene—the age of catastrophes—if survival is to be assured. The traditional meat industry is rapidly depleting natural resources such as land and water.

The growing global population and an increase in average income are fueling both the individual and the total amount of meat consumed in the world.[2] According to the Food and Agriculture Organization of the United Nations, by 2050, annual meat production will increase by 200 million tonnes a year, which is 76 percent more than was produced in 2000.[3] The average meat consumption per person in the world will rise from 39 to 49 kilograms annually, with the highest increase in the developing countries.[4]

Our love for meat might well be the leading cause of the global ecological catastrophe. Animal-sourced foods require more land and use more fresh water than any other human activity. They are also the primary source of greenhouse gas emissions within the food industry.[5]

If the global population cannot sustain itself on meat-heavy Western diets, a switch to eating bugs could bring a promise to guarantee our survival. In recent years, insects have been looked at as a viable protein source for direct human consumption and indirect use as feedstock.[6] This chapter explores *entomophagy*—the practice of eating insects—as a future source of protein. Are insects the best choice for alternative proteins? Will insect farming replace industrial livestock production? And what about the taste? Can insects give us sensory pleasure that is the same as or even better than iconic meat dishes such as salty bacon and juicy, crispy steaks? And is the bug buffet guaranteed to be sustainable?

The History of Insect Cuisine

Entomophagy is nothing new for us. Early humans were omnivorous, and it is believed that insects were an essential component of their diet at least dating back to the Paleolithic era.[7] Why is entomophagy not a significant part of the modern diet? For the world's northern and temperate zones, the answer is heat. The average temperature in these regions is too low to sustain the growth of enough insects or a steady supply of them throughout the year or both. For the first humans who settled in the northern climate zones, eating insects would not have yielded enough nutrition, and with no option to eat them, there was also no reason to put them on the menu. So insects disappeared from our cuisine. Besides the honey bee and silkworm, insect rearing has consequently been excluded from the agricultural revolutions and major innovations in livestock farming. This geocultural explanation is the most credible reason for the absence of insects in the modern food industry.[8]

Today insects have remained an essential part of the diet in only a few cultures around the world. It is estimated that 2 billion people eat insects regularly, with the majority in the Global South.[9] Given the escalating ecological crisis, it is fitting to ask if reintroducing bugs to everyone's daily diet could benefit our future survival.

Ecological Supernutrients

The advantages of farming insects appear to be many. In comparison to industrially produced meat, it is believed that insect protein requires fewer natural resources: land, water, and feed. It also has a much lower carbon footprint.[10] Are insects *the* supernutrient of the future?

A study comparing the rearing of mealworms to other farm animals reveals that the production of one kilogram of mealworm protein requires only 10 percent of the land necessary to provide one kg of beef.[11] Water efficiency is another crucial factor in future agriculture due to the increasing water scarcity. It is estimated that the water footprint for one gram of beef is 112 liters; for one gram of pork, 57 liters; for one gram of poultry, 34 liters; and for one gram of mealworm protein, only 23 liters.[12] And what about the feed? You need only 1.7 kg of feed to produce one kg of crickets compared to 2.5 kg of feed to produce one kg of poultry, 5 kg of feed to produce one kg of pork, and 10 kg of feed to produce one kg of beef.[13] Also, carbon emissions from insect farming are significantly smaller when compared to the production of other animal proteins. One kg of beef produces, on average, 2,850 kg of CO_2, and one kg of pork produces up to 1,130 kg of CO_2. The production of some insect species emits as little as 2 kg of CO_2 per kg of protein.[14]

Insects also have a much higher edibility rate than any other farm animals. While edibility for chickens and pork is 55 percent and for cattle 44 percent, crickets can be edible up to 80 percent. In addition to this high edibility, insects can also have higher energetic values when compared to other animal meats.[15] They are a high-quality protein source containing all the necessary amino acids. The nutritional values, however, differ from species to species and also depend on the metamorphic stage of the insect. Full-grown crickets, for example, have the same protein rate as pork, while the Mopane caterpillar contains 15 grams more protein than pork per 100 grams of the edible portion.[16]

If insects are so much more efficient to produce and provide such a high yield for relatively low environmental costs, why is this traditional food source not reintroduced as a major component of the Anthropocene diet? Why did we lose the attraction to insects as food in modern times?

Overcoming the Taboo

Among the winged insects that go on all fours you may eat those that have jointed legs above their feet, with which to hop on the ground. Of them you may eat: the locust of any kind, the bald locust of any kind, the cricket of any kind, and the grasshopper of any kind. But all other winged insects that have four feet are detestable to you.
—Leviticus, 11:20–23

The Holy Bible is the first written evidence of humans using insects as food. It is also the first indication of entomophagy becoming a taboo with severe eating restrictions.[17] According to Old Testament guidelines, locusts, crickets, and grasshoppers can be eaten, while all other insects are forbidden to consume and will make you dirty if touched.[18] Today, in the Western world, insects are still seen as creepy food. Entomophagy is largely perceived negatively and viewed with feelings of disgust.[19] Even locusts and grasshoppers, mentioned in the Holy Bible, are nowhere to be seen on the menu. Eating insects is associated with primitive behavior from the past or at best as a last-resort emergency food that can be used for survival.[20] Bugs are considered pests and transmitters of disease, not a safe food supply.[21]

As long as eating insects remains taboo in the Northern Hemisphere, the full potential of insects as a global protein source is unlikely to be tested. The influence of Western meat-heavy diets is still growing in emerging nations, fueling a rise in a negative perception of insects as food even in countries where entomophagy has been a long-established tradition.[22]

How can people be persuaded to shift from seeking the globally reliable taste of hamburgers and chicken nuggets to eating all kinds of strange bugs? Luckily, food taboos are known to have been broken in the past. Such was the case of sushi, which was initially perceived with disgust in Western countries but is now an accepted food worldwide.[23] While eating insects is still not a common practice, in recent years, interest has increased, particularly in wealthier parts of society, where people recognize the potential environmental advantages of entomophagy.[24] The growing discussion around the topic means that "the market is ripe for exploring the potential of insects as a food ingredient."[25]

Figure 6.1. The Anthropocene Kitchen, *Ant(i) Pasti*, 2017. Pâté with red wood ants.

How can we make insects appealing as food for those who have never tried to eat them? Is it possible to produce insect food so delicious-looking that a human carnivore drools? How can we package a bag of bugs to make it just as irresistible as a bag of salty and crispy potato chips?

Designing Bugs

Design, packaging, and branding can play significant roles in overcoming the taboo of entomophagy.[26] However, these are formidable tasks. There is a lack of systematic research about the best design approaches to tackle the taboo of eating insects.[27] As entomophagy is not yet customary, widely varying design strategies are applied to marketing.[28] Two distinct approaches for gaining acceptance for eating insects is to market them as either functional foods or exotic foods. The functional approach focuses on the nutritional qualities of insects. As many people are uninclined to eat them, insects are transformed into commonly known food products, such as pasta and protein bars.[29] The second approach promotes eating insects exactly as they are—that is, as a unique, unusual, and exotic food experience.[30]

So can an insect be made to look desirable as food? The now-defunct company SexyFood tried to do just that—make insect eating sexy. The company offered a variety of insect appetizers packed in sleek black and gold metal cans. Each can was numbered so that customers could choose from eleven bug species. The ready-to-eat content was the intact bodies of insects. The can numbered 13 contained water bugs, can 11 came with giant black ants, and can 14 provided whole rhino beetles. The SexyFood merchandise was visually elegant, and the company marketed its insect products as unique and memorable experiences to share among friends and family.[31]

Not a single bug was shown on SexyFood's products. This might be because there is evidence that an image of an insect head on a package can give an impression of a pest, not food.[32] Therefore, hiding insects in a sleek black and gold might be an advantage. At the same time, one challenge with the product was that even if the wrapping was sexy, potential customers were still confronted with a cold serving of a complete, euthanized insect. The elegant packaging therefore presumably could not overcome reservations about eating bugs.

Figure 6.2. *SexyFood*, 2014–2016. Packaging design by Atelier Design (Steven van Boxtel). Insect product packaging.

3D Printed Insect Foods

For many people, putting a full insect with all its texture in their mouths would be a challenge. But if Western consumers show a higher willingness to eat insects as processed food,[33] could they accept insect protein that takes the shape of traditional food items? Another approach could be for insect-based foods to take on entirely new forms, textures, and product types. Just as pasta is produced in various iconic shapes (such as spaghetti, penne, or ravioli with a filling), could insects be developed into similar unique foods? Insects, as any other food ingredient, can take the forms we desire. Here, innovative technologies come in handy.

Such a direction is explored by Susana Soares in the project *Insects Au Gratin* (2011) that reinvents insect cuisine using 3D printing. First, edible insects are dried and ground into a fine powder. Insect flour is then "mixed with icing butter, cream cheese or water, gelling agent and flavouring to form the right consistency to go through the nozzle."[34] Soares uses the insect paste to 3D print unique food objects that can further be either eaten raw or cooked. The result is a series of aesthetical food sculptures that have no resemblance to bugs. By the use of 3D modeling, the final product can take any form desired. While 3D printing is not a new technology, printing insect-based food is a highly novel approach.[35] Combining an unconventional food source with emerging technologies is a potentially good way to introducing insect-based foods to Western consumers. And once we overcome the disgust effect and are willing to buy foods that contain bugs, will we ever enjoy the taste?

The Culinary Properties of Insects

For insects to become part of the Anthropocene diet, it is not enough to make them simply visually appealing. An added bonus would be if this novel food could offer positive sensory and culinary experiences.[36] To gain a foothold on our menus, insects have to fully satisfy our taste buds. So how do they taste? Are insects salty? Sweet? Or spicy? Do they melt on the tongue or crunch between the teeth? The answer might be that they provide all of these taste experiences and more.

There are at least nineteen hundred documented edible insect species,[37] and they offer a wide variety of tastes and textures.[38] Their flavors not only are

Figure 6.3. Original idea by Susana Soares, project coordination by Susana Soares and Andrew Forkes, *Insects Au Gratin*, 2011. 3D printed insect food.

species-specific but also, as with other wild creatures, can be influenced by the environment where they are grown and the feed they are given.[39]

In the book *Creepy Crawly Cuisine: The Gourmet Guide to Edible Insects* (1998), Julieta Ramos-Elorduy reveals that insects can provide almost any taste that we are accustomed to enjoy through other foods.[40] If you would like your meal to taste like meat, try tree worms that have a flavor of pork rinds. For fish flavor, look for aquatic insects such as water boatmen or dragonflies or the eggs of water boatmen, which are known as *Mexican caviar* as they resemble the taste of shrimp. If you wish for plant-based flavors, try termites or wasps that have a nutty taste. Meanwhile, nopal worms can give you the familiar taste of fried potatoes. While the texture varies depending on each species and the stage of their life-cycle, Ramos-Elorduy notes that eating many insects can be compared to the oral satisfaction of eating pretzels or crackers due to their exoskeletons.[41]

Extensive research on the culinary properties of insects has also been done by the Copenhagen-based nonprofit organization Nordic Food Lab. In 2016, Nordic Food Lab launched a documentary film titled *Bugs* and followed it with a book, *On Eating Insects: Essays, Stories and Recipes*.[42] In both projects, the members of Nordic Food Lab share many useful culinary insights about eating and cooking insects from all around the world and their discoveries of many unique tastes. For example, cherry caterpillars actually taste like cherry blossoms, and Amazonian ants have a distinct flavor of lemongrass.[43]

The research by Nordic Food Lab also discovered that insects can be a useful ingredient in gin production. In collaboration with the Cambridge Distillery, the Nordic Food Lab created an alcoholic beverage using red wood ants.[44] Each bottle of its Anty Gin is made using a distillate of sixty-two red wood ants to give the drink citrusy notes. This is achieved thanks to the formic acid ants need for their bodily defense system.[45] If you are not yet ready to taste a full bug, Anty Gin offers the unique experience of tasting a product where insects are only a small part of the production process.

With plenty of cookbooks guiding us through the culinary experience of preparing delicious insect dishes and even beverages, it seems like there is only one question left to answer: how do we source insects? Even harvesting sixty-two

Figure 6.4. Nordic Food Lab and the Cambridge Distillery, *Anty Gin*, 2013. Gin made with red wood ants.

ants sounds like a time-consuming and challenging activity. And for those who live in cities, when was the last time you saw a bug?

Sourcing Insects

Most of the insects eaten in the world are collected in the wild,[46] which means that their availability depends on climate, seasonal changes, and other environmental conditions. Therefore, it can be argued that entomophagy has survived mainly in Africa, Asia, Latin America, and Oceania, due to the optimal climate conditions for insects to thrive in these regions. The tropical climate allows not only for the availability of big insects but also for a steady supply of large insect populations.[47]

In temperate zones, insect availability is seasonal. During winter, there is not much to be harvested. In the wild, only some aquatic insect species can be found on a year-round basis.[48] Their seasonality makes it even more difficult to acknowledge insects as a viable protein source in Western countries.

The wild catch can also differ from year to year. An extreme example is periodical cicadas in the United States, which have developed a unique life cycle of either thirteen or seventeen years. They emerge from underground only at the end of this cycle,[49] with as many as 1.5 million adult cicadas per acre.[50] This supply would allow humans to harvest up to 370 cicadas per single square meter. Even better, there are calendars for upcoming and potential harvests. Because of their unique life cycle, the fifteen registered periodic cicada broods in the United States are precisely mapped based on their location and the next time they are expected to emerge.[51] However, their high density is an important mechanism against predators, and if a brood fails to maintain it, the whole brood can be eliminated.[52] Large-scale wild harvests by humans would irreversibly affect the periodical cicada population. What initially sounds like a careless feast every thirteen or seventeen years could quickly lead into the progress trap by rapidly draining a valuable food source, just as happened with the passenger pigeon population in the 1980s[53] (see chapter 9, Fantastic Cuisine).

Besides the uneven distribution and availability of wild insect species worldwide, ecologists report that many insect populations face a worrying decline. And that is without the global human population chasing after insects as food. It is predicted that 40 percent of the world's insect species will go extinct within decades.[54] Anthropocene-induced challenges such as habitat change, pollution, and climate change are the main drivers for this decline.[55] Is this future protein source disappearing before we even have started to appreciate it in the modern diet?

The consequences of mass extinction could also be something quite opposite. As discussed in chapter 10, Outlook, all previous mass extinctions were characterized by a consequent increase in biodiversity afterward. In the long term, could human-induced climate change result in a new genesis for insects? The global warming of the Anthropocene might extend tropical climate conditions geographically, allowing insects to thrive better and bigger everywhere on the globe. Could such a scenario with an increased population of insects on earth keep all humans well-fed?

Invasive Species and Pests

Even now, in the midst of the Anthropocene extinction, not all insect species are experiencing a population decline. Climate change and human activity, such as global trade, have led to an expansion of so-called invasive species.[56] Invasive species are non-native organisms that adapt to new geographies and locations, where they spread aggressively. They often have adverse environmental effects, threatening native biodiversity, economies, and even human health.[57] A recurring suggestion is to limit the spread of invasive species by harvesting them for consumption.[58] Could we really eat our problems away? What would be the advantage of using invasive insect species as food?

An interesting case is the Asian giant hornet (*Vespa mandarinia*), which is native to Asia and is a feared predator of non-Asian honey bees that are defenseless against its attacks.[59] The Asian giant hornet was detected along the west coast of Canada and the United States in 2020, leading to fears that they would threaten humans and the local honey bee population.[60] While its sting kills many people in Asia every year, the Asian giant hornet has been a traditional source of cheap protein in some rural parts of central Japan.[61] Now it is also served as a delicacy in thirty restaurants in Tokyo.[62] The five-centimeter-long adult insects are said to be delicious and can be prepared in a variety of ways. Their larvae can be marinated to take on the flavor of the liquid, while adults are often fried or used in liquor production.[63] While the Asian giant hornet is pending to be eradicated in North America, why not also harvest it as food?

Several native insects are seen as pests due to their devastating effects on agricultural crops. Why not just eat them? Numerous insect pests are edible and could be controlled through harvesting as food.[64] These include termites, stink bugs,[65] rhinoceros beetles, grasshoppers, and locusts.[66] A well-known pest is the desert locust, currently spreading across eight countries of East Africa. It is considered the most destructive migratory pest in the world as it can form swarms as large as 24,000 square kilometers and travel with the wind up to 150 kilometers per day. During these migrations, desert locusts feed on any green vegetation they can find, including crops. The largest swarms consist of as many as 150 million locusts per square kilometer.[67]

Can we embrace the harvesting of these and other similarly feared insects and use them to strengthen food security? Research reveals that locals in some African regions can earn more money by selling crop pests such as grasshoppers and locusts as food than selling the crop itself.[68] Stink bugs are also agricultural pests that are considered delicacies in some parts of Africa and Asia.[69] If you have pests bugging you in the garden, an obvious solution is to simply serve them for dinner.

Using this collection of pests and invasive insect species for food can increase food security and reduce the need for pesticides. However, it is unlikely that temperate climate zones can provide enough wild insects[70] and even fewer invasive species and pests to feed protein-hungry humans. How then can large-scale insect yields be ensured across the globe?

Insect Farms

If foraging insects in the wild is insufficient to ensure enough protein for the global population, what about rearing them industrially in fully controlled environments? Many researchers see the development of innovative technologies to farm edible insects as "a logical extension of our past and critical aspect of our future success."[71] It is also considered a more sustainable alternative to wild harvesting.[72] As insects can be farmed almost anywhere, they might even be part of a future space diet.[73] However, at present, the diet on earth is what concerns us most.

Particularly pressing is the question of how to feed people living in urban settings. It is estimated that 68 percent of the world's population will live in urban areas by 2050,[74] meaning that most protein will have to be produced for consumption in the cities. If insects can be farmed anywhere with few resources, why not farm them directly where people live? In recent years, several design interventions have explored consumers' willingness to grow insects in their homes.

One such project is *Hive* (2018) by the LIVIN Farms studio, a kitchenware product that allows its users to become microlivestock farmers. With *Hive*, you can breed and feed mealworms (*Tenebrio molitor*) in a well-controlled closed environment for direct human consumption.[75]

Hive is promoted as a sleek design object for a modern household interior. The product envisions insect farming as a simple, kitchen table-top, and easily controllable process—something that anyone can do at home. There might, however, be more problems to insect rearing than seems from the first sight. The *Hive* users have reported incidents such as pest invasions, insect escapes, and even diseases wiping out whole colonies.[76] If you are a beginner in urban insect farming, you might remain with an empty stomach. And even if you are successful, how much protein is it possible to produce through such a kitchen device?

The *Farm 432* (2013) by Katharina Unger provides a concrete answer to the question. *Farm 432* is an earlier version of *Hive* and is named after the number of hours it takes to turn one gram of black soldier fly eggs into 2.4 kilograms of larva protein.[77] If the number of 432 hours is correct, it should be possible to produce approximately forty-nine kg of larva protein per year using *Farm 432*. With the average meat consumption per person expected to reach forty-nine kg annually by the year 2050, household insect farming, in theory, seems to offer a viable alternative. And considering insects' high nutritional value, protein content, and—most likely—low food miles, the same number of insects means much more value than any farm animal meat can provide.

A similar approach to normalizing insect farming through functional design is taken by Icelandic designer Búi Bjarmar Aðalsteinsson in the project *The Fly Factory*. The project investigates how to integrate the production of black soldier fly larvae in Icelandic food culture. Similar to *Farm 432*, *The Fly Factory* is also created as a closed-loop system. In addition to breeding and harvesting chambers, it has an additional compartment for cooling the processed larva products. Such a system reuses the excess heat from the refrigerator to warm up the breeding chamber. At the same time, the waste products from flies serve as nutrients for larvae. In turn, their excrement further serves as fertilizer for growing plants in the breeding chamber. Besides creating a functioning device, the designer also explores potential products that could be conventionalized in Iceland, such as larva paté and larva pudding.[78]

Figure 6.5. LIVIN Farms, *Hive*, 2018.
Mealworms fed with kitchen scraps.

Producing insect-based proteins at home would allow consumers to oversee farming conditions and control the type of feed their insects are given. Insects can also be fed on organic food waste. Such forms of circular reuse could be practiced both in households and in mass-scale insect production, reducing the amount of global food waste.[79]

Instead of stockpiling the food waste as an emergency supply for the future, as proposed in the project *New Earth Meals Ready to Eat (NEMRE)* by Tattfoo Tan (see chapter 2, Ecological Crisis Menu), the 1.2 billion tonnes of food waste every year could be safely used as feed to rear nutritious insects for direct human consumption.

Adding to this, in the Anthropocene, humans also have access to large amounts of nonorganic waste, such as plastics, that could be potentially used as feed for insects. As discussed in chapter 3, Anthropocene Specials, humans cannot directly digest plastics, but several studies have reported on organisms that can, including waxworms, bacteria *Ideonella sakaiensis 201-F6*, and fungi *Schizophyllum commune* and *Pleurotus ostreatus*. Research shows that mealworms are able to survive for more than a month fed solely on Styrofoam.[80] While results indicated reduced productivity and slower growth rates on this diet, the research shows that insects can be used to utilize some of the waste that humans cannot digest.

In the *BoeteBurger Project* by Marc Paulusma (Studio MARC), trash is turned into a treat using Styrofoam-eating mealworms. The designer believes it is important not only to reduce the amount of waste we create but also to address how we deal with existing waste. According to Paulusma, in a world where we could grow our food using the trash we have created, we would build a cycle where "waste is never wasteful."[81] If mealworms can digest polystyrene and break it down on a chemical level, all we would eat is harmless organics. As a validation of concept, Paulusma turns his mealworm colony into an edible insect-protein patty, putting the proof in the pudding.[82]

In the United States alone, seventy-three thousand tonnes of Styrofoam were generated in 2018 and only five thousand tonnes were recycled.[83] Why should we waste such a vast amount of valuable resources if they could be turned into insect protein? Styrofoam and other nonorganic materials that

Figure 6.6. Marc Paulusma, *BoeteBurger Project*, 2019. Mealworms fed a diet of Styrofoam packaging material.

Figure 6.7. Marc Paulusma, *BoeteBurger Project*, 2019. Styrofoam-fed mealworms grounded into a burger.

several organisms are finding taste for is something available in every modern household. Instead of making us feel good through lifestyle ecologism such as sorting trash before disposing it,[84] a large part of nonorganic household waste could be recycled directly into food without leaving the house.

Kitchenware for Eating Insects

With insect farming promising to move directly into the households of future consumers, this novel food source will also come with new traditions and food rituals. Enhancing our experience of entomophagy does not happen only through cooking and sourcing the food. How we put the bugs on the table also matters. With the fork and knife came new eating habits. What innovative tools will affect our eating of insects? And how? The new field of neurogastronomy emphasizes the multimodal impact various stimuli have on the experience of eating. For example, using heavy cutlery is likely to make food more appealing to the palate.[85]

To popularize entomophagy and create new experiences for eating insects, Japanese designer Wataru Kobayashi has created a set of picnic tableware called *BUGBUG* (2016). The project consists of stainless-steel cutlery, including unique chopsticks for picking up insects and specialized serving plates.[86] The project targets families and food enthusiasts of today allowing people to prepare for the large-scale food crisis of tomorrow.

The future cuisine will be defined by ecological disasters, but it also can be rich in traditions and rituals targeting the foods available during the Anthropocene. Instead of focusing on the taste and nutritional value of insects, *BUGBUG* allows us to envision a refined food experience. Creating new traditions when it comes to entomophagy is already possible today. Just as we have special utensils for eating crab, modern eating tools can be reinvented for eating insect cuisine. If products like *BUGBUG* cutlery are introduced directly into the market, consumers of today might find it easier to imagine insects as regular food. New traditions play an important role in the acceptance of novel foods. If we have all the necessary tools to farm, prepare, and eat insects, then how about the food safety of eating bugs? Is entomophagy for everyone?

Figure 6.8. Wataru Kobayashi, *BUGBUG*, 2016.
A textured spoon for grinding insects.

Figure 6.9. Wataru Kobayashi, *BUGBUG*, 2016.
A cutlery set for eating insects.

Food Safety

Food safety standards and regulations play an important role in the food industry. They allow food to be traded within and across nations by ensuring its safety for consumers. Without standards and regulations, novel foods cannot be sold commercially. Therefore, food legislation must also fall in place when it comes to novel foods, such as insects. This is crucial because of the many health hazards that insects might carry.[87] Potential concerns include food allergies, chemical contamination, and biological risks. As an example, individuals with seafood allergies might experience an allergic reaction to eating insects, and people allergic to pollen should definitely not eat honey bees.[88] Reports also show that allergies can be directly triggered by the consumption of insects.[89] And if harvested in the wild, insects can contain pesticides leading to pesticide poisoning.[90] Wild insects also have a higher chance of transmitting parasitic diseases to humans.[91]

These are substantial safety concerns. Perhaps the Bible was not totally wrong with its eating advice. Even if the shift from animal-based protein to insects is an environmentally sound choice, entomophagy entails serious challenges if incorporated permanently and on a large scale in Western diets.[92]

Insects as Vegan Food

For the modern consumer, food choices often come down to what is perceived as more ethically correct. Are there any moral implications in eating insects? With increasing consciousness about the welfare and well-being of nonhuman species, is it ethical to breed and slaughter massive numbers of insects for food? Will vegetarians consume bugs?

A study in Finland concluded that nonvegan vegetarians are most positive toward entomophagy and that vegans consider eating insects irresponsible and morally wrong.[93] These concerns are not groundless. A standard procedure for killing insects for food is to scald them with hot water after one to three days of starvation.[94] It is still unclear if insects can feel pain,[95] but this sounds like a harsh way to treat any living being, even for food.

While livestock rearing is highly regulated in most parts of the world, including Europe, legislation concerning the welfare of farmed insects is mostly absent.[96] Research affirms that environmental factors during the rearing process affect the psychological, behavioral, and ecological conditions of the insect population. What happens to chickens in crowded and feed-restricted poultry farms also happens to insects: a high population density and a lack of food lead to cannibalism.[97] It is believed that legislation regarding farmed insect welfare could positively influence consumer attitudes toward insects as food.[98]

Unsustainable Farming

With an increasingly popular vision of a meatless future and the global need to produce more proteins with fewer resources, insects pose a suitable solution for feeding the world in the Anthropocene. With insect breeding still practiced only on a small scale, it is hard to foresee future challenges if the industry was to expand. What would be the actual environmental impacts of large-scale industrial insect farming?

Large insects do not thrive in temperate climate zones, yet temperate zones are among the places where protein needs to be produced if we are to have a sustainable, short-traveled food supply. Here insect farming has to take place in well-controlled and heated environments. Providing relatively high temperatures for insects results in high energy use.[99] Insect growth time can shift significantly even with a two- or three-degree Celsius difference in temperature.[100] And it is not only about raising the temperature. In temperate countries, the environment has to be both heated during winters and cooled during summers. Besides temperature, the humidity levels and air circulation also have to be just right.[101]

It turns out insects need rather specific climate control to provide proteins. Heating, cooling, air conditioning, and air humidifying take quite a lot of energy. In fact, the large energy demand for insect rearing in temperate climates raises doubts about whether it ever can be sustainable, at least with current methods of insect farming.[102]

So what would be the actual carbon footprint of insect farming in the Northern Hemisphere? What would be the environmental impact if every household added yet another household appliance? Taking into account the expected high amount of energy needed to farm insects in colder climates, will a future bug buffet really have a lower environmental footprint than farmed animals have? This environment-friendly resource might fall into the progress trap just as our love for animal protein did through industrial farming (see chapter 1, Cooking for Survival). A concern here is understanding what happens when a sustainable resource is used in unsustainable ways. One of the authors of *On Eating Insects: Essays, Stories and Recipes*, Joshua Evans, further criticizes the modern trend of simplifying sustainability by noting that "We have gotten into a bad habit of talking about specific organisms as 'sustainable' or 'unsustainable,'" Evans says. "But sustainability is not a property of organisms. It is a property of systems."[103]

If farmed bugs were globally accepted as food, they might eventually share the shortcomings of global-scale insect harvesting in the wild. Just like any other natural resource, insects can also become overexploited and collected using unsustainable methods.[104]

In the journey toward a future bug cuisine, it is important to investigate the potential environmental impact of harvesting and farming insects compared to traditional farming and livestock rearing practices.[105] While such knowledge is still lacking, we might have to simply explore the challenges on the way with some trial bugs on our plate.

Summary

Insects might be the ultimate protein source due to their high nutritional value, fast growth, and low environmental impact—at least in theory. There is also potential to reduce organic waste by using it as feed for breeding insects. And even better, mealworms that can digest Styrofoam can turn nonorganic waste into nutritious food for humans. The wide range of tastes and textures offered by insects make them a valuable food source.

With a long list of advantages, it might seem that the only thing standing in the way of turning this sustainable resource into a global industry is social acceptability, especially in Western countries. There are many approaches that could bring this traditional food back to tables around the world. Design, packaging, and branding can play essential roles. New traditions and rituals can emerge if the market offers specialized tools for harvesting, farming, and eating insects. The food itself can take a more desirable form through the use of 3D printing technologies.

However, even if the taboo of eating bugs is overcome and all regulatory legislation is in place, insects will not be compatible with everyone's diet. There are many potential concerns, from food allergies to chemical contamination, that will repel consumers from eating bugs. And if nothing else, breeding and eating insects might be objected to by the vegan community in the same way that it objects to similar ethical dilemmas in the farm animal industry.

While we are working toward overcoming all these issues, there are more serious questions to answer. Global insect populations are facing a worrying decline, and even without it, insects are not easily sourced in temperate and northern climate zones. This could be solved by farming insects in controlled environments both on an industrial scale and directly in urban households. However, industrial insect farming has never been implemented on a global scale. Therefore, if bugs are ever to become a primary protein source in the world, it will be an experiment on a global scale with uncertain outcomes. Despite our previous achievements of upscaling the farming of animals, it is not a given that we will succeed in scaling up microlivestock production to a global level. The success of small-scale, sustainable insect farming in warmer climates might end quite differently in temperate zones where the need for climate control in itself might render large-scale farms inefficacious and energy-intensive.

With Western countries struggling to appreciate insects as food, the experiment with replacing beef with farmed bugs is likely to be postponed. For now, insects seem to remain on the menu as an alternative emergency food. While the world struggles to swallow bugs, the age-old race for finding the perfect food source continues. For the near future, it seems humanity will need to discover new superfoods other than protein-rich bugs.

Future Superfoods

7 Future Superfoods

Living without hunger, with nutritious and tasty food in abundance, is probably the oldest and most universal dream for *Homo sapiens*. To fulfill it, humans are continually searching for the one and true superfood, sourced through the cornucopia, a horn of plenty that can feed and satisfy everyone around the world. Even better would be a food that has super functions.

An Age-Old Dream

Our fantasies about foods with enormous powers are nothing new. Ancient Greeks told legends about ambrosia, the gods' mythical food that provided immortality to anyone who consumed it.[1] The Greek poet Homer wrote about fruits that were produced without the need for farming. And the ancient Greek comedy *Land of Plenty* portrayed tasty dishes that reproduced by themselves, flowed down the rivers, and fell from the skies with no effort.[2] What a simple life it would be for humans on earth if these fantasies ever came true.

However, modern humans cannot complain about the lack of mythical food experiences in their daily lives. Many of the food realities we live through today would have appeared outright impossible to our early human ancestors and ancient civilizations.

A far-fetched example is *Virtual Shopping Service* by the company LifeStyles in 360 that allows their customers to buy groceries from their own homes using virtual reality goggles.[3] One can only speculate about whether early humans would find this digital 360-degree replica of a local grocery store like a dream come true or a nightmare.

How far have we gotten in our quest for food with ambrosia's powers and for endless food rivers floating from the sky? The year 2016 went down in history as the first time that consumers received their pizzas from a drone delivery service.[4] Yet we have not achieved an endless flow of food. The natural resources on earth are getting more limited by the day. And pizza is also unlikely to be ever perceived as a superfood. So what real superfoods do we already have in our daily diets? And which ones are yet to arrive in the future?

The Perfect Meal

Our perception of what a superfood is changes over time. Since the 1970s, there has been an increased public consciousness of the relationship between our health and the food we eat.[5] Scrinis calls this *food healthism*, pointing to the popular nutrition discourse that has moved beyond the avoidance of bad nutrition and increasingly focuses on the use of good nutrition. Food healthism suggests that instead of simply being healthy, people should follow the pursuit of a perfect diet of foods that "may also be imbued with an exaggerated sense of their ability to cure or prevent a range of health issues, or to enhance the health and performance of the body and mind."[6]

In the past fifty years, nutrition and public health experts have contributed to framing mainstream nutritional advice on how to eat healthily. This advice has also been transformed into a powerful marketing approach for advertising the health benefits of certain foods.[7] The popularity of various healthy eating trends has also come with a heightened interest in superfoods.[8] Because there is no scientific evidence that superfoods are any healthier than their equivalent alternatives not deemed superfoods, it is believed that their popularity is simply triggered by the public perception.[9] An obsession with the perfect diet has become a sweet spot for food manufacturers. It sells. Today the public is exposed to an endless flow of health benefit claims for almost any food available. Some misleading examples are discussed by the National Health Service of the United Kingdom: "curry could save your life," "beetroot can fight dementia," broccoli can "undo diabetes damage," and wine can "add five years to your life."[10]

Paradoxically, the obsession with a perfect and healthy diet can also lead to an eating disorder. Steven Bratman, a practitioner of alternative medicine, refers to it as *orthorexia* or *orthorexia nervosa*.[11] Orthorexia is the unhealthy obsession with having to eat food perceived as being healthy from a subjective point of view. Some so-called healthy foods quite often turn out to be the opposite when people follow trendy but often erroneous dietary advice that does not provide necessary nutrition for everyday needs. The potentially harmful effects of orthorexia are malnutrition, weight loss, fatigue, emotional instability, and even social isolation. These are also characteristics of mental disorders.[12]

To protect consumers against false claims, in 2007, the European Union banned the use of the word *superfood* unless there is substantial research backing a food's health benefits. This ban is further evidence that the superfoods promoted by the commercial sector are more of cultural products than actual foods with enormous powers. What do we mean by the term *superfood*? Should the long-awaited superfood of the future extend the human lifespan like legendary ambrosia? Cure all diseases? Or give eaters full nutritional and taste fulfillment with the least input? Satisfy the appetite for weeks or even months after eating? Can one perfect food source feed the whole world equally? The superfood of the Anthropocene might turn out to be simply the food that can be produced in sufficient quantities despite the scale of ecological disasters.

Cockroach Milk

The search for the ultimate superfood has led to wild scientific studies and speculations about food futures, even if the food in question is of natural origin. A research paper from 2018 speculated that the next superfood might be something quite unexpected—cockroach milk. According to the article, the cockroach species *Diploptera functata* has evolved the trait of viviparity: it can provide nutrients to developing embryos during the gestation period.[13] It produces a substance that contains protein crystals claimed to be more nutritious than any animal milk.[14] Even if cockroach milk was more nutritious than cow's milk, one question remains: how to milk a cockroach?

The scientist and researcher Leonard Chavas claims that the process is energy-intensive and time-consuming.[15] While the method itself is relatively easy, the milk-like substance can be harvested only at the stage of the cockroach's lifespan when it begins to lactate for its offspring. Unlike cows, the cockroach has to be killed in order to extract the milk crystals from its midgut. The article speculates that milking just a couple of cockroaches is likely to take half a day. If the estimation is correct, you would need to kill a thousand cockroaches to obtain only a hundred grams of the product.[16]

Even if cockroach milk is the ultimate superfood and contains four times more protein than cow's milk, it is unlikely to appear on supermarket shelves in the future. While the promise of cockroach milk has taken public debate by storm, in reality, large-scale production is likely to be both unsustainable and insufficient. It might be easier to drink four times more cow's milk to gain the same energy and protein value.

Meal in a Pill

Nature holds many potential superfoods, but in the Anthropocene, sustaining the world solely on foods of natural origin will be just as problematic as feeding everyone with cockroach milk. When nature as we know it changes (at best) or disappears (at worst), a future superfood is likely to be produced synthetically. Many are counting on the technology-driven food futures. Due to food's vital role in the survival of humanity, food speculations are often a recurring theme in the science fiction genre. Science fiction writers concerned with earth's limited resources and the exploding global population usually imagine a world saved by technological inventions and fed in a reductionist manner. Is less more when it comes to food?

The most famous staple food of science fiction is the meal in a pill. This ultimate techno-utopian meal often refers to a small tablet that provides complete nutritional fulfillment. How did the idea of a single food pill become so popular, despite the rich cultural importance of food, cooking, and eating?

According to food researcher and writer Warren James Belasco, our fascination with the meal in a pill goes back to the Neolithic age. Once humans started domesticating grains and legumes, they saw them as minute foods with immense powers. A single seed or a single grain can produce a large amount of food. Therefore, each becomes enormously important. Belasco also draws parallels to charms and gems that in folklore possess magical qualities despite their miniaturization. Similarly, the meal in a pill promises a comfortable and almost magic fix to the most timeless and universal human problem—the struggle for food.[17]

Earliest speculations on the meal in a pill concept date back to the 1880s. For example, in the satirical novel *The Republic of the Future* (1887), American writer Anna Bowman Dodd describes New York in the year 2050, where food tablets are delivered directly to apartments.[18] Dodd explains that the food pills freed women of the future from working in the kitchen: "When the last pie was made into the first pellet, women's true freedom began."[19]

Figure 7.1. Anthropocene Kitchen,
Meal-in-a-Pill, 2017.

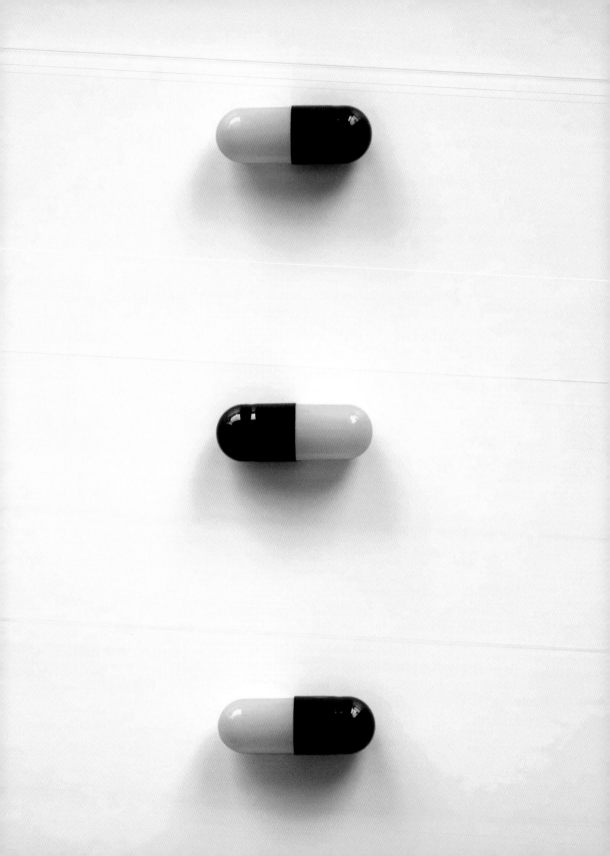

The concept of a meal in a pill did not originate simply out of frustrations with the labor-intensive process of cooking. The science-based claims in favor of this concept were already laid down in the 1890s. In 1894, chemist Marcellin Berthelot predicted that in the year 2000, humans would eat synthetic food in the form of a tablet. He believed that traditional staple foods would no longer be grown naturally but would be manufactured using synthetic chemistry from carbon, hydrogen, oxygen, and nitrogen.[20]

Even if scientific and technological advances in the future would allow us to produce a nutritionally sufficient food pill, could swallowing a single pill be enough to please the taste buds?

In the 1930 musical comedy *Just Imagine*,[21] the protagonist wakes up from a fifty-year coma and finds himself in 1980s New York, where he discovers that all food has been replaced by pills. His first eating experience is a surprisingly complete meal consisting of clam chowder, roast beef, beets, asparagus, pie, and coffee—all compressed in one small capsule. The pill turns out to satisfy hunger and also give a full array of tastes and food textures. After consuming the tablet, the man notes that "the roast beef was a little bit tough."[22]

Since the 1890s, countless claims have been made in favor of food pills. While many foods today are produced using synthetic ingredients, full meal-replacement pills are nowhere in sight. Scientists are still working on a way to fit all the nutritional and caloric needs into one standardized capsule. It is estimated that with current knowledge and technologies, we would have to consume more than 450 pills a day to gain the necessary nutritional intake.[23] For now, a meal in a pill remains a distant dream of rendering mankind independent of nature's resources and potential failures.[24]

Liquid Food Replacements

The closest we have come to materialize the meal in a pill concept is through liquid food replacements such as the products from the Soylent company. The brand's main merchandise is the plant-based powder Soylent, which consists of thirty-nine essential nutrients and promises "a complete meal in less time than it takes to boil water."[25] Most of their customers use Soylent as an occasional meal replacement, but several have experimented with replacing their entire diet with Soylent for extended periods of time. While the effects of fully replacing solid food with liquid food replacements are little studied, solid foods are digested more slowly and, therefore, are also more satiating.[26]

Just like the meal in a pill, the idea of liquid food replacements emerged in close relation to humanity's interest in exploring and colonizing space. Commercially available products similar to Soylent entered the market in the 1960s. An early predecessor of Soylent was Tang—an orange-flavored powdered drink mix that gained popularity in 1962 thanks to American astronaut John Glenn, who became the first American to eat in space. And Tang was part of this historical meal.[27]

Soylent is a plant-based product, but the 120-calorie drink Tang manifested a future of synthetic food as it consisted of less than 2 percent natural ingredients.[28] This fast-food substitute for fresh orange juice was composed mainly of sweeteners and artificial flavorings.[29]

Golden Rice and Other Silver Bullets

While the reductionist approach to feed the world continues to gain its appeal, we more frequently choose to supersize and superupgrade already existing food sources. Genetic modification techniques are often applied in this quest to enhance our favorite staples. The idea of modifying a food source to reach its full potential also leads to a promise of easy fixes and absolute solutions for increasing food security on a global level. The result is highly debated genetically upgraded foods—genetically modified organisms (GMOs)—which are hailed for their enormous potential to increase food security and criticized as magical solutions (that is, silver bullets) to complicated food security issues.

The most controversial and publicly debated silver bullet project is golden rice—a genetically engineered type of Asian rice (*Oryza sativa*) that can biosynthesize beta-carotene that "is metabolized into vitamin A in the human body."[30] The idea to genetically boost the rice plant was born in the 1990s as a solution for ending vitamin A deficiency and minimizing malnutrition, especially in the least developed countries. As rice is regularly consumed by more than half of the global population, it is speculated that golden rice could have a significant impact on food security on a global scale.[31]

The silver bullet approach has also been applied to other plants, including the potato (*Solanum tuberosum L.*). Using techniques similar to those used to produce golden rice, researchers could enhance the potato with provitamin A and vitamin D, resulting in a golden potato. If successful, it could single-handedly increase food security, especially in countries where the potato is already an important staple.[32]

While the debates continue about GMO use in agriculture, we are likely to see "neither high folate rice, nor high iron rice, nor high zinc rice, nor Golden Rice"[33] tested to their full potential as superfoods. And even if these upgraded foods turn out to be less than silver bullets, would it be harmful to have some additional vitamins and nutrients in the food we already eat? And if our existing food sources can be enhanced with the qualities they are naturally lacking, is the only challenge then to produce enough of it?

Supersizing Food

The broiler chicken (as discussed in chapter 3, Anthropocene Specials) has become a major symbol of human abilities to engineer, supersize, mass-produce, and turn a chosen species into a global staple food. Does this mean humans have modified the chicken into a superfood?

The human role in the development of the modern broiler chicken is discussed by artist Andreas Greiner in the work *Monument for the 308: Monument for a Contemporary Dinosaur (Ross 308)* (2016).[34] *Ross 308* confronts the audience with a large sculpture of a broiler chicken in the size of a dinosaur. It is a 3D printed sculpture based on a high-definition computerized tomography (CT) scan of a dead Ross 308 chicken. In comparison to its prehistoric ancestor, the bird-like dinosaur Archaeopteryx, the Ross 308 chicken is not extinct, and it has never been bigger in its individual size or in its popularity in the global population. However, the bird can no longer survive without humans. Therefore, Greiner's monumental work, which resembles a typical dinosaur displayed in a national history museum, "acknowledges the reciprocal dependency between this animal and humans."[35]

The work represents how humans have manipulated the chicken's anatomy to serve increasing levels of global meat consumption. If the modern chicken is more of a genetically engineered product than a naturally occurring superfood, what further transformations will this bird undergo in the future? What will be the next upgrade to our industrially farmed animals?

Redesigning Farm Animals

Once we have upgraded the plants and animals to their full size and nutritional potential, will this finally ensure global food security? As discussed in chapter

Figure 7.2. Andreas Greiner, *Monument for the 308: Monument for a Contemporary Dinosaur (Ross 308)*, 2016. Installation view at Berlinische Galerie.

Figure 7.3. Paul Gong,
The Cow of Tomorrow, 2015.

6, Bug Buffet, as long as industrial farming exists, it has enormous negative environmental effects on carbon emissions and energy, water, and land overuse. If humans are to continue farming animals for meat and milk in the Anthropocene, the industry must be radically reinvented. The answer to sustainable industrial farming in the future might be simply using farm animals for the production of more than just meat and dairy.

In his project *The Cow of Tomorrow* (2015), artist Paul Gong creates a future scenario where already heavily redesigned farm animals can be further modified for energy production. Gong proposes implementing a tiny turbine in the artery of the cow, allowing it to use the blood flow to harness energy. According to Gong, the cow is a perfect candidate due to its long history of industrial domestication. Since we have already genetically modified the cow to enhance its milk yield, why not look into new tasks the animal could possibly do?[36] While *The Cow of Tomorrow* might look like merely thought-provoking speculation, it is inspired by scientific research exploring the technical feasibility of harvesting electric power using a miniature hydrodynamic turbine powered by the cardiac output flow in a peripheral artery.[37]

The artist created the project to look at the question of whether it would be for the better or worse to further increase human control over nature through the use of emerging technologies.[38] The ongoing pursuit of genetically engineered superanimals and superplants brings up many ethical and moral questions. Is there a limit to how far other species should be transformed to benefit human survival? Would creating multipurpose farm animals be unethical if it contributed to sustainability? What would be the environmental benefits of upgrading the cow into a powerhouse? Could the energy harvested from farm animals be used to power the farming industry? In theory, such concepts could help make the resource-intensive industry more self-sufficient. A smart use of existing foods and innovations in biotechnologies can lead to our future superfoods being more than just a source for nutrition.

Unconventional Superfoods: Spirulina

Genetically modified superfoods are not the only foods that humans are skeptical about. Even with the modern trend of superfoods, only a few have become widely popular. How will we know what to choose in the Anthropocene? Many

Figure 7.4. Burton Nitta (Michael Burton and Michiko Nitta), *Near Future Algae Symbiosis Suit: Prototype* (2010).

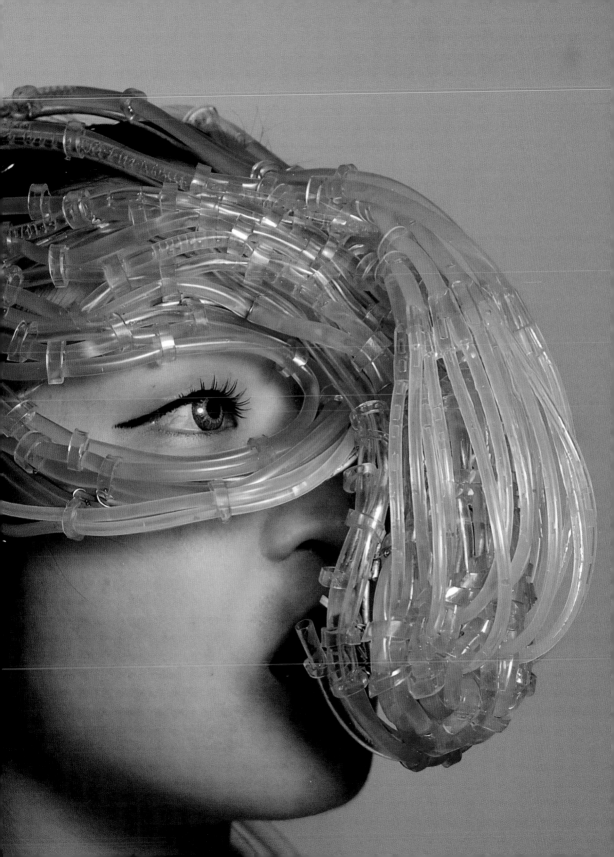

have tried the trending ancient superfoods such as quinoa and amaranth, which was a staple for Aztecs. Also, goji berries, which were used in ancient Chinese medicine, seem to have become a culturally accepted snack. While these foods are unlikely to hold any more nutritional value than other alternatives, there is another ancient superfood that has been praised for its potential in diets after the ecological crisis—algae.

Spirulina platensis (also known as *Arthrospira*) is a genus of cyanobacteria currently gaining global interest as a potential future superfood. Until the sixteenth century, it was used as food by the Aztecs and other Mesoamericans. Today, many African countries still use it as a major protein source by harvesting it from natural waters, drying it, and then eating it.[39]

Dried spirulina holds up to 70 percent proteins, which is half more than soybeans. It also contains various vitamins and minerals and is considered one of the most nutritious foods on the planet. Because of its rich composition, the United Nations World Food Conference declared it in 1974 to be the best food for the future.[40] Spirulina is also likely to withstand uncertain climate futures as it can be cultivated in rather barren areas, does not need clean fresh water, and will gladly grow in highly saline water instead.[41]

Speculation on alternative ways of eating this superfood in the far future is explored by artists Michael Burton and Michiko Nitta in the project *Algaculture* (2010). *Algaculture* is a speculative future scenario that proposes enhancing humans with new bodily organs populated by algae. This unique symbiotic relationship would allow humans to be semiphotosynthetic.[42] As discussed in chapter 2, Ecological Crisis Menu, some organisms are known to enter a coexistence with entities that perform photosynthesis. Could humans possibly follow this blueprint of sustainable and self-sufficient feeding? How would a symbiotic human-spirulina relationship change the way we feed ourselves in the future?

As proposed in *Algaculture*, implementing algae within the human organ system would, in theory, enable humans to carry and produce nutritious food anywhere, even without the need for algae farms. While the project is highly speculative, it also poses the intriguing idea of a closed-loop system based on organic and natural means. Instead of upgrading our foods or developing synthetic food substitutes, we can directly use genetic engineering to transform our bodies and the way we eat (see more examples discussed in chapter 2, Ecological Crisis Menu).

Initially, species of algae look like the perfect superfood for the future, especially in the Anthropocene. However, these organisms are not fully compatible with the human body as food. In 2016, the popular food-replacement company Soylent had to recall products that contained a novel ingredient—*Chlorella protothecoides* algal flour—due to the allergic reaction it caused in several customers.[43]

Several reported side effects raised the alarm about the safety of using microalgae such as spirulina and *Chlorella protothecoides* on a frequent basis. Spirulina is also a cyanobacterium, which has been found to produce toxins. The documented health issues include allergies, nausea, and vomiting. Prolonged consumption of concentrated spirulina in the form of tablets could possibly cause irreversible damage to some organs, including the kidneys and the liver. It is also believed that there are still many compounds that have not yet been discovered and properly tested in microalgae.[44] Choosing the perfect superfood is no easy task, but what if you knew by design exactly what to eat?

DNA-Personalized Superfoods

It might be impossible to produce one standardized superfood providing all the necessary nutritional and medicinal properties to all equally. Instead of looking for just one, could we develop personalized foods that entirely match each individual's needs?

An early inspiration for personalized cuisine can be found in the science fiction television show *Star Trek: The Original Series* (1966–1969), where food was created using a Food Synthesizer—a device that turned different raw materials into edible organic matter. The food synthesizer was able to make a variety of foods with any taste. Further, *Star Trek: The Next Generation* (1987–1994) envisioned the Replicator, an upgraded next-generation device that was preprogrammed to make 4,500 different types of food using a molecular matrix. If it is possible to produce a variety of foods from a few basic ingredients, could we also print superfoods? And if future superfoods are based on data recipes, could we even individualize our food on a personal level?

Can we make the perfect diet for every individual? Currently, we are still in the early stages of understanding the relationship between the human genome, nutrition, and health. Some nutritionists have already begun to use various genomics tests to create individual assessments, and in the future, personalized nutrition could become as common as precision medicine.[45] Companies specializing in genetic testing also offer their clients customized health recommendations based on saliva, blood, and fecal samples. It is now possible to construct meal plans, food supplement packs, and weight-loss recommendations based on individual data.[46] How would the future of the food industry look if ordering the perfect meal was possible through personalized assessment—and if making that meal was as simple as pressing a button on *Star Trek*'s Food Synthesizer?

The Japanese design studio Open Meals explores hyperpersonalization in the food industry through its restaurant concept *Sushi Singularity*.[47] Set to open in Tokyo, *Sushi Singularity* will serve personalized food based on the guests' individual health identification (HID). Prior to the restaurant visit, guests will receive a test kit for obtaining their HID through DNA, urine, and intestinal tests. The restaurant will serve hyperpersonalized sushi dishes fitting each person's individual nutritional needs.

Beyond individual nutritional requirements, the future of personalized superfood could be further customized according to environmental conditions and the availability of local resources.

Summary

Future superfoods will have to fulfill many projections. They should provide health benefits for people, contribute to the survival of the species, enhance our eating experiences, and set a new course for the relationship between humans and nature. Food healthism has led to heightened interest in superfoods and an endless flow of health benefit claims for almost any food. While classic superfoods (such as the meal in a pill) remain a dream of science fiction, new bizarre solutions emerge, but many of them (including cockroach milk) are unlikely to be produced in large enough quantities to impact our diet.

The many upgraded existing foods (such as golden rice, modified to end vitamin A deficiency) show that it is likely that future superfoods will be simply enhanced versions of the staples we know today. The same destiny awaits already upgraded animals such as the broiler chicken. Further, farm animals can be engineered to produce more than just better and bigger meat and dairy. They could even become part of energy production to decrease the carbon footprint of the industry.

Figure 7.5. dentsu / Open Meals, *Sushi Singularity*, 2019.
The health test kit used to customize individual menus.

Figure 7.6. dentsu / Open Meals, *Sushi Singularity*, 2019.
Examples of 3D-printed sushi.

HEALTH TEST KITS

open-meals.com

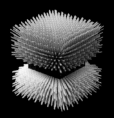 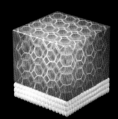 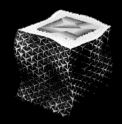 

 SUSHI SINGULARITY 寿超司越

  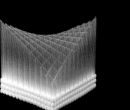 

open-meals.com

Or we can enhance our own bodies with superfoods. By merging the human digestive system with algae, we would enter a symbiotic relationship with one of the most nutritious foods on earth. However, the spirulina super algae have turned out not to be fully compatible with the human body. Humans are prone to allergies and other health risks when consuming spirulina as food.

While it might not be about one single food, the search for the ultimate superfood opens doors to many new and exciting journeys into taste. Emerging technologies also open up possibilities for the creation of personalized superfoods that are fit to individuals based on their DNA and health assessments. Will the future menu consist of only perfected superfoods? If so many superfoods are already available, then why are we not all healthy and eating well? It could be simply that we crave much more than healthy foods.

While some people are overdoing healthy eating and showing symptoms of orthorexia nervosa (a disorder of those who are obsessed with eating healthy), many simply do not bother in the first place. Even if the future might be full of superfoods, humans will likely continue to crave unhealthy snacks. What kind of junk foods will provide our future guilty pleasures?

Future Junk Food

8   Future Junk Food

Junk food is not just what it sounds like. Once it would have been called a superfood. Evolution has given the human body a preference for energy-dense food high in calories. That makes sense historically in the savanna, where food sources were scarce and limited. High-calorie food made survival more likely and life easier. And because junk food is full of fat, salt, and carbohydrates, the body cannot resist it. Evolution made us crave it, and therefore we eat it—a lot. For the first time in human history, high-energy food is readily available on every corner around the globe. It is cheap, too. The growing junk-food industry, however, has dire consequences, affecting human health and physiognomy. Early in this third millennium, very few people live on the savanna anymore. And the rise in junk food has contributed to a quarter of the world's population now being overweight or obese.[1] This has led to an increase in obesity-related diseases such as diabetes and high blood pressure and to environmental devastations caused by resource-intensive agricultural production.

The junk-food industry puts the environment under enormous stress. While it is cheap to produce, highly processed junk food comes at the expense of precious natural resources such as water, energy, and land. And the enormous amounts of fuel needed for both production and transport lead to increased greenhouse gas emissions.

In a world shaped by ongoing ecological crises and resource depletion, it is unlikely that the junk-food industry will remain unaffected by the Anthropocene. If food once again becomes expensive and limited, then perhaps junk food will become a rare and sought-after luxury. Or it might be transformed into something entirely different from what is considered junk food today. That might depend on the contextual definition of *junk*.

What Junk Food?

When is a food called *junk*? The history of junk food dates back to the post–World War II period, when the United States experienced rapid growth in the fast-food industry due to industrialization, the expansion of automobile ownership, and the changing lifestyles of average working-class citizens.[2] Fast-food chains became a cheap and convenient way for American families to eat out of the house at drive-in restaurants. This period was also the beginning of a future obesity epidemic in the United States and beyond.

Michael F. Jacobson coined the term *junk food* in 1972 to define foods high in refined sugar, white flour, fat, salt, and other empty calories.[3] Although high in energy, junk food typically does not offer a complete meal. It is low in several necessary healthy nutrients—usually proteins and vitamins.[4]

While not all fast food is junk food (salads can be fast and yet nutritious), many items on the fast-food menu—such as french fries, hamburgers, chicken nuggets, and pizza—are considered unhealthy when consumed in large quantities and frequently. These foods are also regularly served with sugary carbonated soft drinks, considerably adding to the total of empty energy consumed. These items just taste good, and so it is hard to resist the temptation.

Human Evolution and Junk Food

Human bodies are suckers for junk food. It is simply in our nature. The cravings for energy-dense, sweet, and fatty foods have been essential for human evolution and long-term survival. The consumption of foods high in calories gave our early ancestors higher chances of survival. And the reasons are obvious. It was more efficient to obtain energy-dense foods than energy-poor foods.[5] According to the insurance hypothesis, our ancestors ate when they could, and higher-calorie foods allowed the body to use the energy reserve on days when less food was available.[6]

Today, in a world where energy-dense foods are in abundance, most can obtain the necessary calories and nutrition daily—even all day long, if desired. While the human body is still biologically programmed to collect calories whenever available, the time to use the energy reserve never comes. This leads the excess energy to be stored as blubber in the human body.

Even those with *orthorexia nervosa*—an unhealthy obsession with eating only healthy food—occasionally indulge in erratic episodes of binge eating junk food. For healthy eaters, this is a sin. Such an extreme violation of their dietary rules can lead to "exaggerated emotional distress (fear of disease, anxiety, shame, and negative physical sensations)."[7] Many others simply keep on consuming junk without any concern for their health in the first place. And all of this continues despite endless promises of nutritionally perfected, even personalized superdiets.

Paradoxically, people with orthorexia and people with obsessive junk-food habits have something in common. Both dietary patterns can lead to malnutrition and micronutrient deficiency. If junk food is so full of energy, why is it never enough? Why do humans keep on overeating it? A continued overconsumption of junk foods might be simply the body's way of searching for the nutrients, particularly protein, that junk food usually lacks.[8]

With such a vicious loop of evolution making us want more and more, is there no way out of becoming a junk-food junkie? Or can evolution step in once again to update human eating patterns? If craving energy-dense food is no longer necessary for the well-fed modern human, will future bodies evolve other cravings? In a long-term perspective, it is conceivable that all species will adapt and evolve toward food sources that make them fitter for survival. Currently, this often means leaner and less meat-heavy foods. Likewise, humans might eventually lose the genes driving them to crave sugary and fatty junk foods. Evolution, however, takes time. Can the process be speeded up artificially?

Reprogramming Nature

The consumption of junk food is a severe problem on a global scale. Obesity is no longer an issue only in the Western world.[9] In developed countries, junk food wins over customers by being hyperappealing.[10] It tastes good, and it is cheap. We tend to forget that a healthy lifestyle is not only a recent phenomenon but also one reserved for those with enough means and time. In lower-income parts of the world, junk food provides the most affordable, readily available calories.[11] Undeterred by knowing the potential health hazards and increased environmental threats connected to the production and overconsumption of these foods, many humans continue to indulge themselves in the short-lived pleasures of eating junk.

Junk-food addiction has also reached the animal kingdom. From raccoons scavenging in Illinois[12] to wild macaques terrorizing the city of Lopburi in Thailand,[13] animals have found a taste for it, too. They find it abundantly in urban areas. For them, junk food is their newfound high-energy food source—more of a superfood than anything else. Scientists have observed that the junk-food diet of wild boars in Barcelona allows them to produce larger and more frequent litters. While they grow faster and bigger, they also live shorter lives[14]—just like obese humans.[15]

Are humans just as genetically doomed to binge eat junk food as the wild boars of Barcelona are? If human DNA does not evolve naturally toward new dietary patterns, could it be genetically modified?[16] Is this the only way to reprogram and reengineer the natural cravings for foods?[17] If reengineering is possible, what should be the new perception of junk food? Can the human body be forced to repel unhealthy foods by turning off the urge for overly fatty, sugary, and salty foods? Could people be reprogrammed with a craving for healthy vegetable snacks instead?

Good Junk

Given the abundant availability of junk food, many low-income people around the world can afford a tasty, warm meal at a fast-food restaurant. To proclaim war against all junk food implies that many might not get enough to eat. Junk food is not always bad for those who eat it, and it is not always seen as an obvious hazard. Many iconic junk-food items were once considered good for you and even to have medicinal powers. Digestive cookies were invented by a doctor to aid digestion after a meal.[18] And corn flakes were developed by John Harvey Kellogg to improve the diet of hospital patients.[19]

Even popular beverages now considered highly unhealthy—such as Coca-Cola, Dr. Pepper, and Pepsi-Cola—were created for medical purposes by pharmaceutical professionals operating drug stores.[20] Coca-Cola, the most popular nonalcoholic beverage in the world, was initially created as a medicinal syrup to cure headaches and morphine addiction and sold only in drug stores. The original recipe from 1886 included coca derivatives such as cocaine.[21] And while modern soft drinks do not contain cocaine, their high sugar levels are claimed to be just as addictive.[22] In 2019, 1.9 billion daily servings of Coca-Cola alone were consumed worldwide every day.[23] Like junk foods, junk drinks are a threat to health and the environment.

Water Security Threats

While humans carelessly continue to refresh themselves with carbonated soft drinks, global drinking water reserves are running out. The emerging environmental crises are soon to render accessing fresh water just as problematic as ensuring enough food for the global population.[24] The world's freshwater sources are now being drained faster than they can be replenished through natural processes.[25] The world is under dire water scarcity threats that lead to serious health complications and even fatalities. It is estimated that 829,000 people die each year from diarrhea after drinking unsafe water,[26] and this will increase as the global water demand is expected to grow by 55 percent over the next thirty years.[27]

To raise awareness of water's vital role in human survival and our current unsustainable use of it, artist Helmut Smits takes a critical look at the soft drink industry by naming his project after Coca-Cola's 1970s slogan: *The Real Thing* (2010–2017). Producing one liter of Coca-Cola requires three times as much water. Yet it is the preferred refreshment for many, even if it overexploits already scarce freshwater resources. The really real thing that is needed is not Coca-Cola but the main ingredient used to produce it—water. The artist exposes the reckless production and mindless consumerism of carbonated soft drinks by reversing the manufacturing process of Coca-Cola. For Smits, the real thing is recovered by distilling the beverage back to pure drinking water.[28]

Coca-Cola and other soft drinks are a treat that many find hard to refuse, but what if their access to fresh water was limited? Would people still have the same taste for these drinks? In 2018, the world experienced the most significant water crisis in modern history as the city of Cape Town, South Africa, with its 3.8 million population, despite severe restrictions on water use, was close to running out of water. Shortly before the water taps ran dry for Cape Town inhabitants, also known as *Day Zero*, heavy rains fell and replenished the dams. Yet the scale of long-term water restrictions experienced in Cape Town raises the question of feasibility and sustainability of the soft drink industry from the perspective of long-term water scarcity. Why should scarce fresh water be exchanged for three times less liquid? Is this perhaps another reason to recycle human waste products (urine and feces) back into drinking water (see chapter 5, Human Deli)?

It is only a question of time before the whole junk-food industry will be put in the spotlight of the global environmental crisis. How will it adapt? What new technological developments will take place to keep up with the growing global demand for junk food?

Innovating Junk
The junk-food industry has always been marked by innovation, leading to many notable achievements that are hard to dismiss, despite their evident negative environmental impacts.

The most infamous example might be the invention of the plastic bottle. Polyethylene terephthalate (PET) bottling was born from the idea of storing soda drinks in plastic bottles, and it was patented in 1973 by Nathaniel Wyeth.[29] The use of PET bottles by Pepsi-Cola and Coca-Cola has contributed to the material's immense success while also leading to one of the largest environmental disasters. With more than a million bottles produced every minute,[30] the PET bottle has led to enormous amounts of microplastics in the environment. However, plastic food packaging has also contributed to improved food safety, reduced fuel use for transportation, and extended product shelf life.[31] Adding to the downside of plastic packaging is the large amount of water needed to produce it. An average of five liters of water is required to manufacture a typical single-use plastic water or soda bottle.[32] The water footprint varies widely across countries and individual manufacturing facilities. When calculating the total water footprint that goes into the production of a one-liter PET bottle of drinking water can reach as high as 17.4 liters as documented in India.[33] And the water footprint is even higher when the contents are not just drinking water but soft drinks.

Besides food packaging, the junk-food industry has also played an essential role in the innovation of biotechnologies and genetic engineering. Tyson Foods, for example, is an American company that created an entirely new chicken breed, Mr. McDonald, in the 1980s. Tyson Foods doubled the chicken's size and made it easier to debone just to improve the production process of McDonald's chicken nuggets.[34]

The vast amount of chicken bones in the earth's crust could become one of the geological markers of the Anthropocene (described in chapter 3, Anthropocene Specials). Our love for chicken-based fast foods and the rise of the junk-food industry have directly accelerated the emergence of the Anthropocene as a new geological age.

Today, junk food is more enticing than ever. Since the early 1980s, there have been steady improvements and supersizing of fast food and junk food.[35]

Figure 8.1. Helmut Smits, *The Real Thing*, 2010–2017.
Distilling Coca-Cola back into clean drinking water.

Figure 8.2. Helmut Smits, *The Real Thing*, 2010–2017.

Now the creation of every new junk food is based on science. Each aspect of these unhealthy food experiences is engineered and calculated down to perfection. When developing a new junk-food item, scientists work toward what they call the *bliss point*—the perfect amount of sugar, salt, and fat that keeps the consumer wanting, buying, and eating more.

Junk food has also gotten bigger. By 2007, the portion sizes in fast-food restaurants were two to five times larger than they were when first introduced in the 1950s.[36] While some portion sizes of fast-food meals have been reduced in the last decade, there are always new appealing items on the menu that aim to attract bliss-point-seeking customers. A recent trend has been fast-food and snack-food fusion. From Kit Kat chocolate bars on pizzas to hot dogs wrapped in fried chicken instead of bread buns, these meals combine several junk foods in one dish. These outrageous meals set the taste buds on fire with an explosion of flavors, often a sugar shock, and an unavoidable caloric rush. As with every addiction, after a high comes a low. The consumer is left with withdrawal symptoms and starts looking for an easy fix—which often is simply eating more and more often.[37]

Is the junk food of the future merely going to be bigger and better? And is the global population going to eat more empty calories and grow increasingly oversized? Or are there alternatives to replacing the calorie-dense junk-food staples and their ingredients? That would be good both for the health and the environment.

Healthy Junk

The enormous demand for junk foods and junk drinks is a global challenge that has grown exponentially since the 1950s. However, the emerging trend of seeking healthy diets is increasingly challenging the junk-food industry. Many are optimistic about a transition toward healthy junk food. "It may not be too long before you stop at the fast-food restaurant for a healthful serving of french fries or fried chicken."[38] Some are anticipating an upgrade of the nutritional qualities of already existing junk foods, like Snickers candy bars with added protein and vitamin-infused soft drinks, while others wish to replace all junk food with healthy meals. If humans are naturally programmed to crave fatty and sugary foods, it makes little sense to choose such trendy alternatives as kale chips instead of, well, any junk food that is high in calories. However, in the future, there might be no choice, especially when it comes to meat-based junk foods.

Eight Futures of Meat

The success of the fast-food industry is based largely on industrialized livestock production that provides nearly endless amounts of cheap beef for hamburgers and, more recently, chicken meat for wings and nuggets.[39] An increasing global demand for processed meat has also led to the large negative environmental impacts of industrial farming. China has transformed itself from a country that rationed foods and used grain coupons into a country with a meat-heavy diet.[40] Since 1987, when the first fast-food restaurant (Kentucky Fried Chicken) opened in China,[41] the country's 1.4 billion consumers have gradually shifted toward Western-style diets with high amounts of processed meat. This has led to an increase in obesity in China and other Asian countries.[42]

As discussed in chapter 4, Fake Foodies, finding substitutes for processed animal meat will play an essential role in ensuring that enough protein is produced for the growing global population. It also means that the junk-food meat staples are likely to change drastically in the future. What transformations await our favorite meat snacks?

IKEA's Copenhagen design studio Space10, in its project *Tomorrow's Meatball: A Visual Exploration of Future Food* (2015), envisions a wide range of futures for one of the world's most iconic meat-based dishes—the IKEA meatball.[43] More than 2 million of them are served every day.[44] Space10 addresses how they might look, taste, and be made in the near future and offers eight different scenarios for transforming this well-known staple food.[45] *The Artificial Meatball*, for example, is grown directly in the lab. Biotechnologies have not advanced enough to ensure enough cultured meat to supply IKEA restaurants with 2 million meatballs a day, but (as discussed in chapter 4, Fake Foodies) lab-grown meat holds the potential to radically change the meat industry forever. Another of Space10's eight proposed meatballs that is a more likely near-future protein source is *The Lean Green Algae Ball*. This dish is made from algae protein, a potential future superfood and the fastest-growing plant organism found in nature. Algae have been deemed favorable for future diets due to their high nutritional value and potential to withstand harsh climate conditions. However, as described in chapter 7, Future Superfoods, there are many compatibility issues when it comes to incorporating algae in the human diet.

The Space10 project also proposes an insect-based meatball—*The Crispy Bug Ball*—that could be disguised as the IKEA staple food and finally make bug cuisine available and loved by everyone. If we could succeed in implementing insect menus in sustainable ways and in overcoming the many potential health hazards connected to eating insects (see chapter 6, Bug Buffet), *The Crispy Bug Ball* could become a viable alternative to an animal-based recipe.

And with *The Wonderful Waste Ball*, Space10 utilizes food waste to create a sustainable meal that reduces spoiled food globally. Another concept is *The Urban Farmer's Ball*, which uses city farms to provide a year-round meat-free product with low food miles. According to the project's authors: "Tomorrow's Meatball gets people a little more familiar with the unfamiliar."[46] While *The Artificial Meatball* is unlikely to become a commercially available food item anytime soon, other recipes offered by Space10 are already realizable, if only neophobic consumers would try eating them.

Although more in line with fast food, Space10's *Tomorrow's Meatball* project offers several appealing approaches for producing meat-based junk foods in alternative ways. If we could turn unhealthy and unsustainable products into something good both for human health and the environment, would they still be junk foods? Or would they be staples in a healthy and balanced diet? Another way of looking at future junk foods is to consider ingredients that have fewer calories.

Real Empty Calories

Can junk food become healthier if it contains fewer calories? Would low-calorie junk foods help to decrease global obesity levels? What if junk food was entirely calorie-free? And what should calorie-less junk food be made from? Typical junk foods are highly processed and often contain artificial additives. In some cases, junk-food items are so processed that they could be considered fake foods. They mimic the original dish while being made from alternative ingredients (see chapter 4, Fake Foodies). One example is the popular potato-based snack Pringles, which consists of a mash of flour made from 60 percent wheat, corn, and rice mixed with only 40 percent potatoes. The added salty and rich seasonings might be the only things that distinguish the raw pulp of the snack from tasting industrially neutral and paperlike.

Does it matter what ingredients are used to make junk food as long as it tastes great? The food writer Michael Pollan notes that when we eat the hamburgers, chicken nuggets, and french fries sold in popular fast-food restaurants,

Figure 8.3. Space10, *Tomorrow's Meatball: A Visual Exploration of Future Food*, 2015.

"we forget we're eating an animal or potato."[47] This is a unique effect of many junk-food items. The taste is not associated with the ingredients the food is actually made from. Instead, the unique and universalized flavors are directly connected to the taste of hamburgers and french fries.[48] There is a cognitive and visceral disconnect between the ingredients and the resulting dish.

Could we simply give the same familiar flavoring to alternative foods? When it comes to taste, any food could be turned sweet using the miracle fruit *Synsepalum dulcificum*. Such an approach was discussed by Paul Gong in the project *Human Hyena* (2014), where he turned rotten foods into something delicious for human consumption (described in chapter 2, Ecological Crisis Menu). Can we also turn the nutritionally valueless resources we have an abundance of into something tasty?

HAF Studio has created a humorous take on such a food scenario in the project *Slim Chips* (2010). *Slim Chips* are Pringles taken to their maximum artificiality by being rendered with the main ingredients of edible paper, organic food coloring, and flavoring.[49] You can choose from the tastes of peppermint, blueberry, and sweet potato. These zero-calorie paper treats promise an experience compared to eating "tasty air" with no weight gain.[50] Because edible paper is usually made from starches and sugars, *Slim Chips* are likely to contain some calories. But a zero-calories recipe is possible if actual cellulose fibers (like those used in the production of real paper) are added. Cellulose fibers pass through the digestive system as unabsorbable waste. Until the human body is modified to gain full nutritional benefits from eating cellulose—as suggested in the Gints Gabrans project *FOOOD* (2014) (see chapter 2, Ecological Crisis Menu)—a chewable and yet real paper would be the perfect fit for calorie-free junk-food snacks. An indigestible future junk food might be a healthy fit for a world gone cataclysmic. Then again, in a nominalist twist, junk food should be made from junk.

Junk Food from Real Junk

To entirely rethink the concept of junk food, why not consider the waste products of the Anthropocene as potential ingredients? What if it was possible to utilize the real junk of the age of ecological catastrophes in food production? Can nonorganic global waste become food? Such a food scenario is presented by designer Marc Paulusma in his *BoeteBurger Project* (2019), which turns trash

Figure 8.4. Henry Hargreaves and Caitlin Levin, *Deep-Fried Gadgets*, 2013.

into a treat by using Styrofoam-eating mealworms (see chapter 6, Bug Buffet). Can all human-made waste become food one day?

Artist Boo Chapple in her project *Consumables* (2009) imagines a time when all manufactured electronic devices are edible. Instead of throwing out an old phone when a new model arrives in the market, it could be eaten. How would this affect global food security? Chapple playfully speculates that the global food crisis can be solved by turning e-waste into e-food. Collecting old phones could even become a way for the poor to supplement their daily diet.

In Chapple's vision, phones should not only be edible but also taste good, and they should come in different flavors like bubblegum and banana.[51] The project envisions a whole new cultural shift toward the idea of edible electronics. These edibles would also come with new food traditions and rituals. The new diet would lead to new cultural items, such as a series of cookbooks on how to cook your phone,[52] adding to the playfulness of the project.

Why stop at eating only phones? Photographer and artist Henry Hargreaves further explores an electronics-infused cuisine in his project *Deep-Fried Gadgets* (2013) in collaboration with food stylist Caitlin Levin. The photography series of *Deep-Fried Gadgets* explore similarities between tech and fast-food cultures by visualizing ready-to-eat dishes of deep-fried earplugs, laptops, smartphones, and other electronic devices.[53]

Boo Chapple's *Consumables* and Henry Hargreaves and Caitlin Levin's *Deep-Fried Gadgets* were created as critical commentaries on a modern throwaway culture that often offers solutions to the problems of consumerism by encouraging more consumerism.[54] However, these projects also draw attention to the fact that massive amounts of human-made waste have no use in nature or the global food chain. As discussed in chapter 3, Anthropocene Specials, the new raw ingredients of the Anthropocene, such as plastics, might become a useful resource in the future. There might even be ways to eat these hyperobjects. While several organisms on earth can already digest plastic, if electronics are to become food, they have to be produced as edible in the first place (at least for now). If that is the case, utilizing e-junk to produce junk food is a thought-provoking and compelling thought exercise about our food futures.

Summary

Whatever future junk food is made of, it probably will be just as high on our menu as junk food is today. Modern junk-food items are perfected to the bliss point, that ideal combo of the three irresistible food qualities—sweet, salty, and fatty. These qualities make people want to eat endlessly more. And the consequences are dire. The global population is getting more obese, while the reckless production of junk foods is making a severe dent both in the environment and in natural resources.

Although the trend toward healthier diets is increasing, a global and collective shift toward eradicating junk foods entirely is unlikely. Humans are not genetically programmed to drool at green low-calorie vegetables the same way they do over fatty hamburgers and sugary beverages. For now and until human nature is reengineered to eat differently, junk-food obsessions are here to stay.

However, the age of the Anthropocene is unfolding, and the junk-food industry is bound to be affected by the coming environmental crises. To satisfy a growing consumption of junk foods when natural resources are depleted, these foods will soon have to be made using alternative ingredients. When anything can be made to taste like anything else, a circular and sustainable approach would be to produce junk food using waste that is already available in abundance—from food waste to paper and even plastics. Alternative production methods for meat-based junk foods will be a particular focus for researchers, as processed meat is one of the main pillars of the unsustainable fast-food industry. One suggestion is to replace animal-based meat with insects fed on nonorganic waste materials such as Styrofoam.

Modern junk foods have succeeded in satisfying the natural cravings of humans. Nonetheless, junk food is much more than just ingredients. It is also the food fantasies that many of these foods are built around. The Unicorn Frappuccino by Starbucks is one such example.[55] This mythical and sugary beverage is topped with a colorful mix of sweet pink and blue sour powder and looks like something in a fantasy book. The food futures of the Anthropocene might be likewise magic. What if the onset of ecological catastrophes could bring about even more fantastic meals?

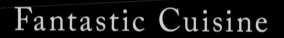

Fantastic Cuisine

9   Fantastic Cuisine

Anthropocene cuisine should do more than satisfy the nutritional needs of humans. It must offer meals that make the wildest food fantasies come true. However, envisioning fantastic food futures can be challenging when crisis seems to be the default mode of the age we live in. How then are we to expand the upcoming cuisine beyond imagination and realize impossible meals? With the emerging scale of global environmental catastrophes, are fantastic food futures even achievable? This chapter takes a deeper look at some wild attempts to expand the human diet into radically new and previously unimaginable menus. During the barren years, innovative thinking will be our plenty. Now, at the onset of the epoch, we are starting to see the first signs of a future cuisine that will come with new flavors from far away, rediscovered foods from the past, and dishes that have existed only in our imagination.

Fantasizing Foods

Food fantasies are nothing new. Humans have dreamt of a food paradise since ancient times.[1] As described in chapter 7, Superfoods, ancient Greeks craved ambrosia—the mythical food that provides immortality to humans and gods alike. The Nordic Vikings dreamed about arriving at the warriors' heaven, Valhalla, where there was plenty of food thanks to the phoenix boar called Sæhrímnir, which would be slaughtered and eaten every night, only to be brought back to life again the next day for another feast. And if the pig's meat would make you thirsty, no worries. On the roof of Valhalla Hall stood the goat Heidrun, whose udders provided a steady flow of mead, satisfying even the thirstiest immortals.

More recent fantasies, as imagined by science fiction writers, offer less eternity but simpler eating—the complete meal in a pill, which provides a fully nutritional and tasteful food journey contained within a little capsule (see chapter 6, Superfoods).

Today, humans often fantasize about indulging themselves in existing foods, often hyperappealing junk foods, such as the creatively marketed Unicorn Frappuccino by Starbucks (see chapter 8, Junk Foods). Other modern food fantasies focus on pursuing a diet of natural and organic foods (see chapter 1, Cooking for Survival). In the Anthropocene—the age of ecological catastrophes—these fantasies might be just as utopian as producing food that provides immortality.

Nonetheless, food fantasies are an essential tool for speculating about our edible futures. What is not possible today could be possible tomorrow. Our future menu will be shaped by the ideas and meals we think about today. To ensure fantastic food futures, our food dreams deserve to be more than just efficient eating with full nutritional benefits. How do we think big and wild in an age marked by the unfolding sixth mass extinction of species?

A De-extinction Menu

Even challenges connected to the Anthropocene extinction and changing environmental conditions can be turned into food fantasies. There are numerous urban legends about and actual instances of eating already extinct animals. One of the most famous occurred in 1951, when the Explorers Club served a meal made with centuries-old mammoth meat found frozen in Alaska.[2] The Explorers Club, which had an annual tradition of serving exotic and rare dishes, initially labeled the meat as an American giant ground sloth, also known as *Megatherium*. However, the story quickly evolved into a tale of feasting on a mammoth. In 2016, a specimen from the meal preserved in the Yale Peabody Museum was DNA sequenced to confirm its origins, and it turned out to be neither. The served meat was from the green sea turtle instead.[3] While the story was demystified fifty-five years after the meal, the fascination with the supposed dish persists. Many might be curious to taste a historic and iconic animal like a mammoth. In the future, this could become a reality. And unique dishes that only early human ancestors have eaten could become part of the next human diet.

Figure 9.1. The Center for Genomic Gastronomy, *The De-extinction Deli*, 2013.

Figure 9.2. The Center for Genomic Gastronomy, *The De-extinction Deli*, 2013.

Emerging biotechnologies such as breeding back, cross-species cloning, and genetic engineering come with a promise that extinct species could be brought back to life.[4] Reversal of extinction, also known as de-extinction, can be defined as "the process of creating an organism which is—or greatly resembles—a member of an extinct species."[5] De-extinction has become a popular topic both in public debates and in the laboratories of scientists. First efforts have already been made to recover the woolly mammoth, the flightless bird moa, the Chinese river dolphin, and the famous passenger pigeon, among others.[6] The concept of de-extinction challenges humans to reassess their relationship with nature. It also inspires the gastronomic imagination. What will happen when the de-extinction process becomes a reality? Will we bring back extinct species in the future just to eat them again? What meal from the past would you like to serve for dinner? Could a juicy steak of *Tyrannosaurus rex* become the new favorite dish at high-end restaurants? Or could it be something anyone can afford as a fast-food meal? While science is still evolving to accomplish such life-, death-, and evolution-redefining acts, art can offer a glimpse into these possibly fantastic gastronomic futures.

The De-extinction Deli (2013) by The Center for Genomic Gastronomy is an art project that speculates about how emerging biotechnologies might bring back extinct species to include them in our diet.[7] With the slogan "Yesterday's Meat Tomorrow," the project is realized in the form of a market stand that questions the risks and outcomes of de-extinction if it becomes a reality.[8]

The various interactive materials presented in the project include a woolly mammoth cut chart, a sample of the passenger pigeon's habitat, and even a grow-your-own-cells kit. These objects ask if we would be willing to eat the de-extinct and how such newly introduced species would change the current habitats.

The De-extinction Deli raises many environmental, ethical, cultural, and purely gastronomic questions, allowing for wild speculations about our future diet. While the project debates the motivation for bringing the extinct back to life, it also envisions unintended consequences that might emerge along the way.[9]

What might be the ethical dilemmas of making de-extinction possible in the future? Would global food security increase if large animals like mammoths and dinosaurs were part of the daily protein source? Would prehistoric animals be considered invasive species even if some of them lived on earth long before modern-day animals? And how would the attitude toward the preservation of any nonhuman species change if it was possible to diminish their numbers and resurrect them on demand? If a species could be brought back to life by human choice, some would be favored, and others would be neglected. Just as humans pay attention to a few chosen endangered species and work toward the recovery of their populations, the same probably would be true when it comes to de-extinction. Dinosaurs roaming the earth in the twenty-first century is such an appealing picture that they have appeared in many movies, including the Jurassic Park movie franchise by Steven Spielberg. Meanwhile, the Steller's sea cow population that was annihilated by hungry Russian hunters in 1768[10] is rarely mentioned (see chapter 3, Anthropocene Specials, for a discussion about mass extinctions). The ethical dilemmas of the future might ask if we are obligated to bring back all the species that have gone extinct due to direct human actions—that is, the many we have eaten to extinction.

The modern broiler chicken might be compared to something of "a contemporary dinosaur" (see chapter 6, Future Superfoods), but eating a modern chicken is not the same as eating a real dinosaur. So the work on how to bring extinct species back to life continues. For now, the human imagination has to be satisfied with such imposter foods as Dino Buddies—dinosaur-shaped chicken breast nuggets.[11] If bringing back entire animals is not yet possible, can we grow fantastic beasts in the lab instead?

Fantasies from Realities

New exotic foods in the future are unlikely to be of natural origin. They are more likely to be cooked in the lab. While de-extinction is not yet a reality, many other biotechnologies allow for high levels of experimentation with living organisms.

As discussed in chapter 4, Fake Foodies, artist Kuang-Yi Ku proposes creating new mythical creatures in the *Tiger Penis Project* (2018). He theorizes that it could be possible not only to grow individual organs in the lab but also to merge DNA from several creatures to create entirely new ones. Ku proposes developing lab-grown foods that use the DNA of animals that are believed to have healing powers in cultural myths. If we can already culture animals, fish, and even human meat in the lab, growing new foods by mixing several DNA will likely be feasible soon. With such speculation on the table, the possibilities for entirely new foods are endless.

These achievements in biotechnologies allow the science-based imagination to blossom. What if we could create and eat creatures that have never existed on earth? It might even be possible to combine DNA from a horse and a human to make a lab-grown centaur—the creature from Greek mythology that has the upper body of a human and the lower body of a horse. What about the human fish? What would such a DNA-blended mermaid taste like? Mythical creatures like mermaids and unicorns have so far lived only in people's imaginations. Having them pop up on a plate near you would trigger both interest and new ethical dilemmas.

Edible Myths

De-extinction methods are underway, but humans have an uneasy relationship with many endangered species, being trapped between craving and conscience. Did the Russian sailors know that they might be hunting and eating the last Steller's sea cow? And was the last living passenger pigeon in the wild consumed by a hungry farmer or by a wealthy diner at a high-end restaurant (see chapter 3, Anthropocene Specials)? While these individuals probably did not know the significance of their meals, endangered species are much more closely observed today, making it possible to acknowledge and trace the few that remain. This knowledge is foreseeably contributing to a new food fetish—eating the last of its kind. This fetish follows the law of supply and demand. The fewer of a species, the higher its value and the demand for it. And the value is most often found in its edibility. Such a fetishizing could soon be the case with the Atlantic bluefin tuna—a species that is seriously overexploited. It might not be long until the Atlantic bluefin tuna faces abrupt collapse and extinction. Likewise, there are fewer than a hundred individual Amur leopards left in the wild.[12] Eating a species into mythology did not happen only with the mammoths. How would we deal with mythical creatures appearing on our menu for the first time?

A meal defined by both imagination and extinction was at the center of the performative food experiment *Anthropocene Feast: Eating the Last Dragon*

Figure 9.3. Anthropocene Kitchen, *Anthropocene Feast: Eating the Last Dragon*, 2017.

Figure 9.4. Anthropocene Kitchen, *Anthropocene Feast: Eating the Last Dragon*, 2017.

(2017) by artist collective Anthropocene Kitchen.[13] At the feast, project organizers oversaw the public cooking and serving of what they called the last dragon on planet earth. The event was framed as a celebration that was open to the public, where participants were offered pieces of meat from the barbeque of a three-meter-long dragon meat sculpture. It was a challenging experience both for old and young. During the event, a thought-provoking question was raised by a little boy. As he was chewing his first bite, he asked his mother: "Mommy, why is the dragon real?"[14] According to common knowledge, dragons exist only in myths and fantasies, but this boy's question opens up for further speculation on the question of why dragons cannot be real. Could we possibly eat dragons in the future? What if our fantasies would trigger not just our imaginations but also our palates? The next dragon served might become not the last of its kind but the first to ever exist. So instead of focusing on the tastes of the past through de-extinction, we might eat new species for the first time.

Fantastic Spices

Setting taste to food is an ancient art. Strange seasonings from far away set fire to taste buds in Medieval Europe, and the spice trade changed the globe and fueled the first worldwide trade routes. Tastes can come from places even farther away, such as meteorites falling to earth. A type of cosmic mudball meteorite that landed in Costa Rica has been reported to have "a distinctive aroma that is somewhat like that of cooked Brussels sprouts."[15] This makes for an interesting question: can extraterrestrial objects be eaten and have useful culinary properties? While it was reported that the 1.8-kilogram meteorite dubbed *Aguas Zarcas* had a unique food-like odor to it, geologists supposedly did not taste it due to the possibility that it contained harmful materials.[16]

What previously untested flavors can humans now source thanks to new technologies? Some of the far-sourced foods tasted by humans include moon dust. Sampled by NASA astronauts during the Apollo 17 mission in 1972, it is said to have tasted of gunpowder.[17] A new generation of space exploration might pave the way for extraterrestrial cuisine. Bringing new and never before tasted flavors from around the universe might also inspire meals based on our imaginings about other organisms that might be living far far away.

Eating the E.T.

The human imagination can lead us into new and uncharted edible territories. If a taste for extraterrestrial foods is developed, then future cuisines might allow us to fantasize about eating organisms from outer space. Such a scenario is explored in the project *Eating E.T.—Mock Alien BBQ* (2014) by the Sweden-based design and innovation studio Unsworn in collaboration with artist Terje Östling. The project authors made a life-size gluten replica of the famous science fiction character from the movie *E.T. the Extra-Terrestrial*.[18]

The first public E.T. barbecue took place at the symposium *Exploring the Animal Turn* at the Pufendorf Institute in Lund, Sweden. The fictional character was roasted on a spit and served to symposium participants.

According to one of the project authors, Erik Sandelin, the project *Eating E.T.—Mock Alien BBQ* plays with "our intimate relations to other species, real and fictional."[19] The project asks why we choose to eat some species but consider it unethical to consume others. As the extraterrestrial meal was made from seitan (wheat gluten), many commentaries by the visitors debated meat-mimicry practices. Other ethical considerations asked about the place of humans in the hypothetical intergalactic food chain: "What if aliens came here and they were more intelligent than us. Would they have the right to eat us, as we eat animals?"[20] After discussing the important place the human body already has as food for other organisms (described in chapter 5, Human Deli), we need to ask about our future role as food beyond earth's boundaries and for organisms that do not yet exist.

Sandelin explains that the experience of creating, grilling, and eating the fictional character from outer space engages observers both emotionally and experientially. Both the creators of the project and the participants who had to decide whether to eat the extraterrestrial's body were affected by the experience. The project's team members found themselves taking extra care with its body and arranging the bound arms in what seemed to be a respectful position. Many guests found the experience discomforting and commented that "It is jarring to see his charred body on the spit."[21]

Eating E.T.—Mock Alien BBQ illustrates the complex and often uneasy relationship that exists between humans and their food. Few people would have any discomfort when roasting farm animal meat, but in this project, eating a

fictional character that was never real in the first place becomes more personal than eating a grilled chicken. Does this mean fantastic meals can lead to a human diet that is more considerate of other nonhuman species?

Summary

It is unlikely that the most fantastic food items in the future will be provided directly by Mother Nature. Future meals might contain strange, even extraterrestrial flavors to spice up the food made from the depleting resources on earth. Our soon-to-be cuisine might come from radically different resources as new technologies let us source and create new and mythical foods. Emerging biotechnologies such as genetic engineering, cellular agriculture, and de-extinction open up a vast amount of yet unexplored possibilities for food production. In the future, we might bring back legendary foods from the past. Our menu might consist of dinosaurs, mammoths, passenger pigeons, and other species that we once ate to extinction. This will come with new ethical dilemmas about which species, if not all, to resurrect and how to include them in present habitats without allowing them to become invasive. We might even enter a paradoxical loop and bring them back just to eat them into extinction over and over again. Beyond the obsession with foods from the past, we might extract DNA from several species to create new mythical ones. Results could lead to lab-grown centaur and mermaid meat. The myths and fantasies of eating unicorns, dragons, and extraterrestrial creatures might become an everyday thing in the future. Anthropocene cuisine can be full of extraordinary meals.

As we fast-forward with humanity's constant chase for better and tastier meals, the ecological catastrophes are bound to taste delicious. Whatever the future might hold, our imagination is what pushes the food industry and innovation ahead. Failure in solving a crisis is a failure of imaginations.

Figure 9.5. Terje Östling and Unsworn Industries, *Eating E.T.—Mock Alien BBQ*, 2014.

Figure 9.6. Terje Östling and Unsworn Industries, *Eating E.T.—Mock Alien BBQ*, 2014.

Outlook

10 Outlook

Bringing back extinct species on our plates? Eating aliens? Growing human meat in vitro? You might ask what these activities have to do with securing the survival of the human species. Will consuming smog cookies and bugs fed on Styrofoam really help save human civilization? Can edibles made from human materials actually provide a significant number of nutritious meals?

The recipes presented throughout *The Anthropocene Cookbook* are only part of the answer to how we can feed a growing human population in the future. There are no step-by-step guidelines for future food making. A meal is more than the sum of its ingredients. Shaping our future cuisine requires understanding the time and context in which the food is sourced, prepared, and served. Coming to grips with the environmental conditions of the Anthropocene is necessary because those conditions trickle down to our future survival and food production. So by themselves, tasting smog and feeding bugs on plastic will not save 10 billion people worldwide. Rather, as examples of liminal thinking and problem solving in our liminal times, these projects promote an understanding of the bigger picture of our soon-to-be foodscapes.

Trying to understand the extent of the current and coming global environmental crises can be overwhelming. If a failure to solve a crisis is a failure of imagination, then perhaps we can survive and even thrive if we can get our hands on the appropriate toolboxes. Our survival kit for the future needs to consist of new guides and tools of liminal thinking that our ancestors did not use or need.

Sourcing food today is nothing like it was for early humans. We no longer roam the savannah looking for easy prey. Hunting all mammoths and all passenger pigeons to extinction was a tempting pitfall of the past. Going forward, we must walk a fine line between utilizing new resources and being lured into progress traps. The end of food as we know it is arriving, and we could help ensure that it is not the end of humanity. While the Malthusian trap has not yet come to pass, the lessons learned from many collapsed societies in the past can be of use as we adapt to changing environmental conditions and thereby avoid the bleak finales of Mayan, Norse Greenland, Easter Island, and Aztec civilizations. One essential lesson is that people must eat according to what is available in their *oikos* (Greek for "household" or "place to live" and the root for *economics* and *ecology*)—in their own households and in their future home, the Anthropocene.

The Anthropocene is marked by foodscapes of extreme complexity and infinite unknowns. While traditional resources are depleting, innovative thinking and emerging technologies can open up boundless possibilities of novel food production. Because we cannot predict food availability in the times to come, we cannot ensure our future survival by providing practical solutions and concrete menus of foods to eat and foods to avoid at present. Instead, the works described in this book exemplify the liminal thinking needed to approach extreme future food scenarios and black swan events. Surviving the ecological crises of the Anthropocene will be no easy task. So we must keep all options on the table. Speculative concepts and thought-provoking artistic experiments serve as valuable assets for envisioning previously unthinkable menus and exercising our capacity for liminal thinking until the tide turns.

Today, the earth is catering a decadent feast for humans. We are decimating the garden of Mother Nature by draining off its water and turning its soils to dust. At the far and barren end of this blissful time of abundance, we are likely to take down nonhuman species with us along the way. What, then, will we do if the Anthropocene extinction is imminent? Surely that must be the end?

A New Genesis

The extinctions of species and the human impact on the earth's biodiversity are usually perceived to have only negative effects. What if, on the contrary, these events also open up brand-new possibilities for the interactions, adaptations, and even evolution of nonhuman species?

What if biodiversity increases due to human-made changes in the Anthropocene? While many species are currently going extinct, others are surviving and spreading into new parts of the world where they could not have traveled without the help of humans.[1]

Chris D. Thomas argues that apart from the emergence of new habitats, entirely new species are coming into existence. Examples include the apple fly, which emerged in North America when fruit trees were imported from Europe, and a new bird species, the Italian sparrow, which evolved from the house sparrow and the Mediterranean Spanish sparrow. According to Thomas, the current rate at which new species are coming into existence might be the highest ever. What if we are on the verge of Genesis number six?[2] All previously known five mass extinctions have been characterized by a consequent increase in biodiversity afterward. Therefore, in a million years from now, earth could be supporting more species than ever. They might be different from the ones we know today, but they could be more diverse than ever—all because humans interacted with the environment.[3] It might be too late for *Homo sapiens*, but for life itself, there are signs that it will continue even after the end. Until then, we still have a large population of humans to supply with food.

The Anthropocene Is Now

Genesis number six is in the far future. Or it might never arrive. Until then and probably forever, every generation will have to continually reassess how it produces food and adapt to whatever crises might arise. Providing food for the future cannot be just a recycling of tasty, time-tested recipes. If humans continue to daydream about the historically bountiful and ecological foods supplied by the garden of Mother Nature, survival will indeed be uncertain. Acknowledging the new Anthropocene foodscape, artificial and constructed as it may be, is a first step to take toward creating a recipe for our future continuity.

Paradoxically, even if we wanted to avoid eating Anthropocene-specific ingredients—such as plastics, human-made radiation, and genetically modified foods—we would fail. Hyperobjects such as microplastics are in the seafood we consume, in the water we drink, and even in the air we breathe. We live in the Anthropocene, and all its human-made resources unavoidably find their way into our diet, sooner or later. How we deal with these conditions comes down to choice. We can ignore their arrival and blissfully set a course toward our demise. Or we can utilize them as an opportunity to feed ourselves better, healthier, and tastier than before. Following the recipes for liminal thinking, the first step is to imagine.

Future Eating
Eating choices have always been culturally defined, so thinking about what we should eat next involves keeping track of evolving intricate dietary patterns—healthy foods with no nutritional value, unhealthy foods with the necessary calories for a whole week, emerging eating disorders, healthism, and veganism. There will never be just one universal diet. Humans are omnivores, and their food choices are exactly that—choices. Neomnivores are eating only lab-grown meat, possibly grown from their own DNA, but they provide just a glimpse of the ethical dilemmas facing us in the future when the new food sources discussed in this book become part of the daily diet. With our diet set to become more diverse, the future will come with a number of new, strange eating patterns.

In the future, food phobias and philias are likely to intensify due to the many potential novel foods. Our neophile nature will continue to drive food innovations, and our persistent neophobia toward new foods will keep the bad decisions at bay. Both will be handy as we surf the coming food crisis. Ecological catastrophes are here to stay. And so the Anthropocene becomes our kitchen.

The food concepts presented in this book provide hints about what our next cuisine may be. Following this guide to thinking, future generations will have much more to eat than just the leftovers of today. While many new foods might be organic (such as de-extinct animals and lab-grown food), many upcoming dishes will contain ingredients that our ancestors did not have in their diet. Such radical changes in human dietary patterns might even lead to the next step in human evolution. If including meat in the diet of early human ancestors changed their physiology and physiognomy, then the new ingredients in the future Anthropocene diet are likely to act in similar ways on our bodies. Like in Paul Gong's *Human Hyena* project (see chapter 2, Ecological Crisis Menu), the human body might evolve to eat differently, digest new foods, and even evolve new taste modalities. The taste buds of our ancestors evolved to warn them about potentially dangerous and poisonous foods. Future humans might require additional gustatory modalities specific to the new raw ingredients of the Anthropocene. What if the next generations can precisely evaluate the amount of pollution and even microplastics in food directly through taste? Newly available resources in the Anthropocene and new ways for producing food promise to take human evolution further.

What if we make the wrong food choices? Can our diet lead to the devolution of humans? If meat has played an important role in human brain evolution (see chapter 1, Eating for Survival), what would happen if everyone on earth became vegetarians or even vegans? One open question is how the Anthropocene diets described throughout this book could affect the human body and brain in the long term. Health is important, but omnivore and scavenging humankind has survived until now despite what we ate rather than because what we ate was pure and of high quality. As the projects described in the book have shown, art is food for thought but can also be a way of stirring up the palate in the real world.

Cooking in Real-World Contexts

If, according to Slavoj Žižek, "Most of what we think of as radical or subversive—or even simply ethical—doesn't actually change anything,"[4] are the many speculative ideas discussed in this cookbook substantial enough to bring about real changes?

Trying to predict the future in liminal times is a slippery task. So how can art accomplish real change? The more than sixty creative projects cited in this book manifest the kind of liminal thinking that is required to approach seemingly hopeless circumstances and engage with the most inextricable problems (see the introduction). Actual change in these liminal times starts with a shift in thinking but stands in a close relationship with our palate and with many hands-on, creative endeavors in practical cooking.

What happens when speculative ideations and hypothetical proposals find their way into the real world? Several projects discussed throughout this book have transgressed from art and critical design practices into commercial products and lab experiments or have served as influential incentives for innovation in the food industry.

The in vitro meat experiments by Oron Catts and Ionat Zurr (see chapter 4, Fake Foodies), for example, have served both as a critical conversation piece and as an inspiration for further experiments in scientific contexts and labs. This ripples into the emergent alternative-meat industry. The viability of lab-grown meat products is being tested by start-up companies and even restaurants.[5] There is also a new wave of lab-grown meat experiments speculating further about real-life applications, such as cultivating human meat or animal organs from threatened species to be used in traditional Chinese medicine (see chapter 4, Fake Foodies).

Another example is the independent research and design lab Space10, which created eight alternative future prototypes of the iconic IKEA meatball in 2015 (see chapter 8, Future Junk Food). While most of the project's ideas could be deemed as too speculative to directly be launched through the company's food chain, they challenged IKEA to produce a better veggie meatball to its customers.[6]

A third example is the LIVIN Farms studio (see chapter 6, Bug Buffet), which has commercialized its *Hive* project into an insect farming kit. It is both an educational and a practical tool to start breeding and eating insects at home.

These three examples show how art and designerly projects can affect and influence food production beginning today. The speculative concepts explored in this book hint at the fantastically possible food realities of tomorrow. It does not stop here. The explorations toward our next foods have only just begun, and the future suddenly offers more prospects than limitations.

The Future Is an Empty Page

This book was written in the years of plenty as a call for action against an arid future. While a fantastic food future might be hard to envision when total crisis is the default mode of perception, ecological catastrophes can taste fantastic if we design and compose their organoleptic qualities to be so. If any readers find *The Anthropocene Cookbook* to be pessimistic about the current state of the earth, they should know that even darker predictions were made in former times.

The Nuremberg Chronicle by Hartmann Schedel, published in 1493, was a remarkably rich illustrated book about world history.[7] Toward the end of the book were three empty pages, which future owners of the book could use to document how the world developed during their own lifetimes. After those three empty pages, Schedel illustrated his predictions for the future that he thought was coming, ending with the doomsday that is predicted in the Bible. For many, the end has always been near. As Julian Cribb has written:

As they draw closer they begin to feel its power.
Closer and they can at last make out what it is, this fount of all the ruin,
the suffering, the hunger, the loss. It's a cookbook.[8]

And for those who are terrified by the prospects of the Anthropocene and its associated effects, such as Hothouse Earth, we have good news. The current geologic age will end soon. At the latest, it will end with the onset of the next glacial period. After all, we are living and ruling in an interglacial—a brief warm period on earth between ice ages. The normal appearance of earth is not blue and green but bright, ice cold white sprinkled with a few blue dots. What we will eat then will require an entirely different cookbook.

Now let's get wet.

Acknowledgments

This book has been cooking for more than six years. The idea was born out of an interest in cookbooks as powerful cultural phenomena and the Anthropocene as an important ingredient for future thinking about food.

The Anthropocene Cookbook would not have been possible to write without all the artists, designers, researchers, and thinkers whose work we investigated and have presented herein. Our deepest thanks go first and foremost to them and their contributions to the use of food as a medium for art, thinking, and shaping futures. We are also deeply grateful to them for providing the many images included in this book.

We would like to thank Noah J. Springer, our inspiring editor at the MIT Press, for taking on this project and its extensive and often challenging research materials. Further, our thanks go to the anonymous reviewers at the MIT Press, whose supportive and stimulating feedback encouraged us to continue our work. We also thank the other scientific commentators at the MIT Press, whose expertise provided an analytical assessment of the book's materials from a scientific point of view. Thanks also to our encouraging and precise production editor, Deborah Cantor-Adams, and manuscript editor, Rosemary Winfield, at the MIT Press, as well as Lea Stenslie, University of Amsterdam, for her careful proofreading.

We would not have been able to publish this book without the generous support from TEKS–Trondheim Electronic Arts Centre in Trondheim, Norway, and its director Espen Gangvik. In particular we thank him for believing in the project and promoting the ideas within the book and liminal thinking for the future.

We thank all the people who provided opportunities for us to present our research. We value their critical comments and the comments of the people who participated in numerous events and conversations. Many topics in the book were discovered through practical experiments with novel foods and as a result of passionate discussions with members of our audiences.

An early version of the project was introduced to a wider audience at the Twenty-third International Symposium on Electronic Art (ISEA 2017): Bio-Creation and Peace, in Manizales, Colombia, in June 2017. This was the first of many exciting encounters we had with members of the general public and experts from the fields of art, design, and technology. In the same year, we also had an opportunity to conduct field research in Manaus, Brazil, as part of LABVERDE: Art Immersion Program in the Amazon. Thanks go to Ieva Ūbele and Kaspars Goba, who invited us for a residency at the Rucka Art Foundation in Cesis, Latvia, where over the summer of 2017 we organized a series of experimental and edible lectures. As part of our food adventures in Latvia, we have to thank many people, especially chefs Juris Latišenoks and Jānis Sproģis, who helped us to realize various future food concepts for the public events. We are also grateful to Andra Magone, who made it possible to organize an Anthropocene celebration and feast at the iconic Cesis Castle. It was also an honor to be invited by producer Loreta Bērziņa to discuss the book live on Latvian National Radio. Special thanks go to Vytautas Michelkevičius for inviting us to present our research at the Vilnius Academy of Arts in Vilnius, Lithuania, in 2017 and for welcoming us to a residency program at the Nida Art Colony in Nida, Lithuania. The project was later presented at the Eighth Inter-format Symposium on Rites and Terrabytes, which was held at Nida in June 2018. We also thank the Norwegian BioArt Arena (NOBA), where we presented our ongoing research for the book. Ursula Münster, director of the Oslo School of Environmental Humanities (OSEH), gave us a wonderful opportunity to talk at Nordic Environments: Opening Conference of the Oslo School of Environmental Humanities in November 2019. These events have been important opportunities for extended discussions about the future of food and about ecological thinking at large.

Those who gave feedback throughout the writing process deserve a big thanks for their great help and assistance: Sunčica Ostoić from Kontejner in Zagreb, Croatia; Hege Tapio from i/o/lab–Centre for Future Art in Stavanger, Norway; the philosopher and writer Einar Øverenget; and Lea Stenslie. We are particularly grateful to Alessandro Ludovico from *Neural* magazine, who engaged with us in discussions about the text and also helped us navigate the world of publishing. We also thank Jussi Parikka for his assistance and publishing advice. Further, our supporters and inspirations include Boris Debackere from V2_Lab for the Unstable Media in Rotterdam, the Netherlands; Jurij Krpan from the Kapelica Gallery in Ljubljana, Slovenia; and Nina Czegledy, adjunct professor at Ontario College of Art University, Toronto.

Notes

Introduction

1.   Paul Virilio, "Surfing the Accident," interview by Andreas Ruby, in *The Art of the Accident: Merging of Art, Architecture and Media Technology*, ed. Bart Lootsma, Joke Brouwer, and Arjen Mulder (Rotterdam: NAI Publishers and V2_Organization, 1998), 40.

2.   Will Steffen, Paul J. Crutzen, and John R. McNeill, "The Anthropocene: Are Humans Now Overwhelming the Great Forces of Nature?," *AMBIO: A Journal of the Human Environment* 36, no. 8 (2007): 614–621.

3.   Julian Cribb, *The Coming Famine: The Global Food Crisis and What We Can Do to Avoid It* (Berkeley: University of California Press, 2010), 187.

4.   Filippo Tommaso Marinetti, *The Futurist Cookbook*, ed. Lesley Chamberlain, trans. Suzanne Brill (London: Penguin Books, 2014; first published in Italian in 1932 as *La Cucina Futurista*).

5.   United Nations Educational, Scientific, and Cultural Organization (UNESCO), "Futures Literacy: An Essential Competency for the 21st Century," accessed July 29, 2021, https://en.unesco .org/futuresliteracy.

6.   Theo Reeves-Evison, *Ethics of Contemporary Art: In the Shadow of Transgression* (New York: Bloomsbury Visual Arts, 2020).

7.   Ståle Stenslie, *Virtual Touch: A Study of the Use and Experience of Touch in Artistic, Multimodal and Computer-Based Environments* (Oslo: Oslo School of Architecture and Design, 2010), 204.

8.   Matt Malpass, *Critical Design in Context: History, Theory, and Practice* (London: Bloomsbury Publishing, 2019).

9.   Bruce M. Tharp and Stephanie M. Tharp, *Discursive Design: Critical, Speculative, and Alternative Things* (Cambridge, MA: MIT Press, 2019), 8.

10.   Anthony Dunne and Fiona Raby, *Speculative Everything: Design, Fiction, and Social Dreaming* (Cambridge, MA: MIT Press, 2013).

11.   Nassim Nicholas Taleb, *The Black Swan: The Impact of the Highly Improbable* (New York: Random House, 2007).

12.   Giorgio de Chirico, "Meditations of a Painter: What the Painting of the Future Might Be, 1912," trans. Louis Bourgeois and Robert Goldwater, in *Theories of Modern Art: A Source Book of Artists and Critics*, ed. Herschel B. Chipp (Berkeley: University of California Press, 1996), 397.

13.   Boris Debackere et al., *Innovation in Extreme Scenarios* (Rotterdam: V2_ Institute for the Unstable Media, 2014).

14.   Jean Anthelme Brillat-Savarin, *The Physiology of Taste: Or Meditations on Transcendental Gastronomy*, trans. M. F. K. Fisher (New York: Everyman's Library, 2009; first published in 1949 in New York by Heritage Press), 15 (citations refer to the Heritage Press edition).

Chapter 1

1.   Warren James Belasco, *Meals to Come: A History of the Future of Food* (Berkeley: University of California Press, 2006), 62; John Boyd Orr and David Lubbock, *The White Man's Dilemma*, 2nd ed. (London: George Allen & Unwin, 1965, first published 1953).

2.   DK, *The Story of Food: An Illustrated History of Everything We Eat*, foreword by Giles Coren (New York: DK Publishing, 2018), 9.

3.   DK, *The Story of Food*, 9.

4.   DK, *The Story of Food*, 10.

5.   Laure Ségurel and Céline Bon, "On the Evolution of Lactase Persistence in Humans," *Annual Review of Genomics and Human Genetics* 18, no. 1 (2017): 297–319.

6.   Pascale Gerbault et al., "Evolution of Lactase Persistence: An Example of Human Niche Construction," *Philosophical Transactions of the Royal Society B: Biological Sciences* 366, no. 1566 (2011): 863–877.

7.   Michael Symons, *A History of Cooks and Cooking* (Urbana: University of Illinois Press, 2004; first published in 1945), 104.

8.   Yuval Noah Harari, *Homo Deus: A Brief History of Tomorrow* (New York: Harper Perennial, 2016), 23.

9.   Thomas R. Malthus, *An Essay on the Principle of Population, as It Affects the Future Improvement of Society; with Remarks on the Speculations of W. Godwin, M. Condorcet and Other Writers* (Bellingham, WA: Electronic Scholarly Publishing Project, 1998; first published in 1798 in London by J. Johnson).

10.   Cribb, *The Coming Famine*, 155, referring to Jared Diamond, *Collapse: How Societies Choose to Fail or Survive* (New York: Viking, 2005).

11.   Damian Carrington, "Sixth Mass Extinction of Wildlife Also Threatens Global Food Supplies," *The Guardian*, September 26, 2017.

12.   Dave Goulson, "The Insect Apocalypse, and Why It Matters," *Current Biology* 29, no. 19 (2019): 967–971.

13.   Alexandra-Maria Klein et al., "Importance of Pollinators in Changing Landscapes for World Crops," *Proceedings of the Royal Society B: Biological Sciences* 274, no. 1608 (2006): 303–313.

14.   Ronald Wright, "Ronald Wright: Can We Still Dodge the Progress Trap?," The Tyee, September 20, 2019, https://thetyee.ca/Analysis/2019/09/20/Ronald-Wright-Can-We-Dodge-Progress-Trap/.

15.   Ronald Wright, *A Short History of Progress* (Toronto: House of Anansi Press, 2011; first published in 2004), 8.

16.   S. T. Turvey and C. L. Risley, "Modelling the Extinction of Steller's Sea Cow," *Biology Letters* 2, no. 1 (2006): 94–97.

17. Andrea Dell'Apa, M. Chad Smith, and Mahealani Y. Kaneshiro-Pineiro, "The Influence of Culture on the International Management of Shark Finning," *Environmental Management* 54, no. 2 (2014): 151–161.

18. Mort Rosenblum and Mar Cabra, "In Mackerel's Plunder, Hints of Epic Fish Collapse," *New York Times*, January 25, 2012.

19. Selina M. Stead and Lindsey Laird, eds., *Handbook of Salmon Farming* (London: Springer, 2002); Stephen Hume et al., *A Stain upon the Sea: West Coast Salmon Farming* (Madeira Park, BC: Harbour Publishing, 2004).

20. Wright, *A Short History of Progress.*

21. Wright, *A Short History of Progress.*

22. Wright, *A Short History of Progress.*

23. Evan D. G. Fraser, "Social Vulnerability and Ecological Fragility: Building Bridges between Social and Natural Sciences Using the Irish Potato Famine as a Case Study," *Conservation Ecology* 7, no. 2 (2003).

24. Joe Hasell and Max Roser, "Famines," Our World in Data, last modified December 7, 2017, https://ourworldindata.org/famines.

25. Hasell and Roser, "Famines."

26. Alex de Waal, *Mass Starvation the History and Future of Famine* (Cambridge: Polity Press, 2018).

27. Mark Joseph Stern, "What Causes Famine?," Slate, April 16, 2014, https://slate.com/technology/2014/04/climate-change-and-agriculture-global-warming-could-cause-catastrophic-famines.html.

28. Koert van Mensvoort and Hendrik-Jan Grievink, eds., *Next Nature: Nature Changes Along with Us* (Amsterdam: Next Nature Network, 2015), 273.

29. Serpil Aday and Mehmet Seckin Aday, "Impact of Covid-19 on the Food Supply Chain," *Food Quality and Safety* 4, no. 4 (2020): 167–180.

30. Chris Weller, "China Is Facing a Food Problem: And the Western Diet Could Be to Blame," World Economic Forum, June 19, 2017, https://www.weforum.org/agenda/2017/06/china-is-facing-a-food-problem-and-the-western-diet-could-be-to-blame.

31. Steven Bratman, "Orthorexia vs. Theories of Healthy Eating," *Eating and Weight Disorders* 22 (2017): 381–385.

32. Bratman, "Orthorexia," 383.

33. Toby Miller, *Greenwashing Culture* (Abington, UK: Routledge, 2018).

34. Slavoj Žižek, "Slavoj Žižek: 'Humanity Is OK, but 99% of People Are Boring Idiots,'" interview by Decca Aitkenhead, *The Guardian*, June 10, 2012.

35. Žižek, "Humanity Is OK."

36. Timothy Morton, *Ecology without Nature: Rethinking Environmental Aesthetics* (Cambridge, MA: Harvard University Press, 2007), 205.

37. Van Mensvoort and Grievink, *Next Nature*, 283.

38. Morton, *Ecology without Nature.*

39.   United Nations, "Growing at a Slower Pace, World Population Is Expected to Reach 9.7 Billion in 2050 and Could Peak at Nearly 11 Billion around 2100," June 17, 2019, https://www.un.org/development/desa/en/news/population/world-population-prospects-2019.html.

40.   Michiel Korthals, *Before Dinner: Philosophy and Ethics of Food*, trans. Frans Kooymans (Dordrecht, NL: Springer, 2004), 4.

41.   Korthals, *Before Dinner*, 99.

42.   Richard Fleischer, dir., *Soylent Green*, 1973.

43.   Kevin Ashton, *How to Fly a Horse: The Secret History of Creation, Invention and Discovery* (New York: Anchor Books, 2015), 88.

44.   Andrew F. Smith, *Potato: A Global History* (London: Reaktion Books, 2011).

45.   Ashton, *How to Fly a Horse*, 85.

Chapter 2

1.   Bo Gräslund and Neil Price, "Twilight of the Gods? The 'Dust Veil Event' of AD 536 in Critical Perspective," *Antiquity* 86, no. 332 (2012): 428–443.

2.   Gräslund and Price, "Twilight of the Gods?"

3.   Food and Agriculture Organization of the United Nations (FAO), *Climate Change and Food Security: Risks and Responses* (Rome: Food and Agriculture Organization of the United Nations, 2015).

4.   Nassim Nicholas Taleb, *The Black Swan: The Impact of the Highly Improbable* (New York: Random House, 2007).

5.   Julian Cribb, *The Coming Famine: The Global Food Crisis and What We Can Do to Avoid It* (Berkeley: University of California Press, 2010), 77.

6.   Seth D. Baum et al., "Resilience to Global Food Supply Catastrophes," *Environment Systems and Decisions* 35, no. 2 (2015): 301–313.

7.   William R. Cotton and Roger A. Pielke, "Nuclear Winter," in *Human Impacts on Weather and Climate*, 2nd ed. (Cambridge: Cambridge University Press, 2008), 203–219.

8.   Gillen D'Arcy Wood, *Tambora: The Eruption That Changed the World* (Princeton, NJ: Princeton University Press, 2014), 2.

9.   Smithsonian Institution, "Current Eruptions," Global Volcanism Program, National Museum of Natural History, accessed August 16, 2021, https://volcano.si.edu/gvp_currenteruptions.cfm.

10.   Genesis 41:26–31.

11.   Baum et al., "Resilience to Global Food Supply Catastrophes," 302.

12.   Jonatan A. Lassa et al., "Revisiting Emergency Food Reserve Policy and Practice under Disaster and Extreme Climate Events," *International Journal of Disaster Risk Science* 10 (2019): 1.

13.   Lassa et al., "Revisiting Emergency Food Reserve Policy," 1.

14.   Chris Ellis, "The Noah Virus: Who Is Infected with High Resiliency for Disaster?" (unpublished manuscript, last modified March 15, 2021), 4, Microsoft Word file.

15.   Ellis, "The Noah Virus."

16.   G. Wanat, "Poland: Maintaining New Eating Habits after Covid-19 2021," Statista, September 17, 2021, https://www.statista.com/statistics/1264052/poland-maintaining-new-eating-habits-after -covid-19.

17.   Jenny Gustavsson et al., *Global Food Losses and Food Waste: Extent, Causes and Prevention* (Rome: Food and Agriculture Organization of the United Nations, 2011), v–vi.

18.   Monika van Den Bos Verma et al., "Consumers Discard a Lot More Food than Widely Believed: Estimates of Global Food Waste Using an Energy Gap Approach and Affluence Elasticity of Food Waste," *PLoS One* 15, no. 2 (2020): 1–14.

19.   Tattfoo Tan, *New Earth Meal Ready to Eat*, 2013, accessed July 24, 2020, http://www.tattfoo .com/new_earth/NewEarthMRE.html.

20.   Baum et al., "Resilience to Global Food Supply Catastrophes."

21.   David Charles Denkenberger and Joshua Pearce, *Feeding Everyone No Matter What: Managing Food Security after Global Catastrophe* (London: Academic Press, 2015), 32.

22.   Cribb, *The Coming Famine*, 194.

23.   Denkenberger and Pearce, *Feeding Everyone*, 28.

24.   Baum et al., "Resilience to Global Food Supply Catastrophes."

25.   Baum et al., "Resilience to Global Food Supply Catastrophes," 305.

26.   Jelle Bruinsma, "The Resources Outlook: By How Much Do Land, Water and Crop Yields Need to Increase by 2050?," in *Looking Ahead in World Food and Agriculture: Perspectives to 2050*, ed. Piero Conforti (Rome: Food and Agriculture Organization of the United Nations, 2011), 3.

27.   Kurt Benke and Bruce Tomkins, "Future Food-Production Systems: Vertical Farming and Controlled-Environment Agriculture," *Sustainability: Science, Practice and Policy* 13, no. 1 (2017): 14.

28.   Jimmy Tang, *Future Food Hack: An Experiment with the Future of Food*, 2015, accessed July 24, 2020, http://jimmydesign.co/work/future-food-hack.

29.   Kathrin Specht et al., "How Will We Eat and Produce in the Cities of the Future? From Edible Insects to Vertical Farming—A Study on the Perception and Acceptability of New Approaches," *Sustainability* 11, no. 4315 (2019): 1–22.

30.   United Nations, Department of Economic and Social Affairs, Population Division, *World Urbanization Prospects: The 2018 Revision (ST/ESA/SER.A/420)* (New York: United Nations, 2019), 10.

31.   Jon Henley, "The Future of Food: Inside the World's Largest Urban Farm—Built on a Rooftop," *The Guardian*, July 8, 2020.

32.   Henley, "The Future of Food."

33.   Henley, "The Future of Food"; Marielle Dubbeling and Henk de Zeeuw, "Urban Agriculture and Climate Change Adaptation: Ensuring Food Security through Adaptation," in *Resilient Cities: Cities and Adaptation to Climate Change—Proceedings of the Global Forum 2010*, ed. Konrad Otto-Zimmermann (Dordrecht: Springer, 2011), 443.

34.   Mazhar H. Tunio et al., "Potato Production in Aeroponics: An Emerging Food Growing System in Sustainable Agriculture Forfood Security," *Chilean Journal of Agricultural Research* 80, no. 1 (2020): 118–132.

35. Tunio et al., "Potato Production in Aeroponics."

36. Specht et al., "How Will We Eat and Produce?," 9.

37. Specht et al., "How Will We Eat and Produce?," 13.

38. Baum et al., "Resilience to Global Food Supply Catastrophes"; Denkenberger and Pearce, *Feeding Everyone*, 17–22.

39. Baum et al., "Resilience to Global Food Supply Catastrophes."

40. Baum et al., "Resilience to Global Food Supply Catastrophes," 306; Denkenberger and Pearce, *Feeding Everyone*, 65–66.

41. Julian Chela-Flores, "Terrestrial Microbes as Candidates for Survival on Mars and Europa," in *Journey to Diverse Microbial Worlds: Adaptation to Exotic Environments*, ed. Joseph Seckbach (Dordrecht: Springer, 2000), 387–398.

42. Ernest Geiger, "Problems Connected with the Possible Use of Plankton for Human Nutrition," *American Journal of Clinical Nutrition* 6, no. 4 (1958): 394.

43. R. J. Blakemore, "Nature Article to Commemorate Charles Darwin's Birthday on 12th February," VermEcology, February 12, 2017, accessed December 23, 2019, https://vermecology.wordpress .com/2017/02/12/nature-article-to-commemorate-charles-darwins-birthday-on-12th-feb.

44. Zhejun Sun and Hao Jiang, "Nutritive Evaluation of Earthworms as Human Food," in *Future Foods*, ed. Heimo Juhani Mikkola (Rijeka, Croatia: IntechOpen, 2017), 127–141.

45. Leon C. Megginson, "Lessons from Europe for American Business," *Southwestern Social Science Quarterly* 44, no. 1 (1963): 4.

46. Paul Gong, *Human Hyena*, 2014, accessed July 24, 2020, https://www.paulgong.co.uk/Human -Hyena.

47. Lars Östlund et al., "Bark-Peeling, Food Stress and Tree Spirits: The Use of Pine Inner Bark for Food in Scandinavia and North America," *Journal of Ethnobiology* 29, no. 1 (2009): 104.

48. Anna Sigrithur and Avery MacGuire, *Tree Bark*, Nordic Food Lab, November 24, 2015, https:// nordicfoodlab.org/blog/2015/11/tree-bark/.

49. Björn Corander, *Barkbröd*, accessed January 14, 2021, http://www.kolumbus.fi/bjorn.corander/ bakning.htm (site discontinued); Sigrithur and MacGuire, *Tree Bark*.

50. Sigrithur and MacGuire, *Tree Bark*.

51. Gints Gabrans, *FOOOD*, 2014, accessed July 24, 2020, https://www.gabrans.com/foood.php.

52. Rusty Blazenhoff, "An Edible Desert Survival Manual by Land Rover," Laughing Squid, May 10, 2012, https://laughingsquid.com/an-edible-desert-survival-manual-by-land-rover/.

53. Tomas Linder, "Making the Case for Edible Microorganisms as an Integral Part of a More Sustainable and Resilient Food Production System," *Food Security* 11, no. 2 (2019): 265–278.

54. Jani Sillman et al., "Bacterial Protein for Food and Feed Generated via Renewable Energy and Direct Air Capture of CO2: Can It Reduce Land and Water Use?," *Global Food Security* 22 (2019): 25–32.

55. Sillman et al., "Bacterial Protein for Food," 26.

56. Solar Foods. "Solein® Protein out of Thin Air," accessed August 24, 2021, https://solarfoods.fi/solein.

57. Jean Christophe Valmalette et al., "Light-Induced Electron Transfer and ATP Synthesis in a Carotene Synthesizing Insect," *Scientific Reports* 2, no. 1 (2012): 2.

58. Mary E. Rumpho, Elizabeth J. Summer, and James R. Manhart, "Solar-Powered Sea Slugs: Mollusc/Algal Chloroplast Symbiosis," *Plant Physiology* 123, no. 1 (2000): 29.

59. Michael Houellebecq, *The Possibility of an Island*, trans. Gavin Bowd (2005) (New York: Vintage Books, 2005).

60. Maddie Stone, "Eating the Sun: Can Humans Be Hacked to Do Photosynthesis?," Vice, February 10, 2015, https://www.vice.com/en/article/3dk4bv/human-photosynthesis-will-people-ever-be-able-to-eat-sunlight.

61. Michael Burton and Michiko Nitta, *Algaculture*, accessed August 16, 2021, http://www.burton nitta.co.uk/Algaculture.html.

Chapter 3

1. H. K. Williams, "Religious Education," *The Biblical World* 53, no. 1 (1919): 81.

2. Federico Zuolo, Chiara Testino, and Emanuela Ceva, "The Challenges of Dietary Pluralism," in *The Routledge Handbook of Food Ethics*, ed. Mary Rawlinson and Caleb Ward (Abington, UK: Routledge, 2017), 94.

3. Carys E. Bennett et al., "The Broiler Chicken as a Signal of a Human Reconfigured Biosphere," *Royal Society Open Science* 5, no. 12 (2018): 180325.

4. Tara Garnett, "Plating Up Solutions," *Science* 353, no. 6305 (2016): 1202–1204.

5. A. S. Bawa and K. R. Anilakumar, "Genetically Modified Foods: Safety, Risks and Public Concerns—A Review," *Journal of Food Science and Technology* 50, no. 6 (2013): 1035–1046.

6. Will Steffen, Paul J. Crutzen, and John R. McNeill, "The Anthropocene: Are Humans Now Overwhelming the Great Forces of Nature?," *AMBIO: A Journal of the Human Environment* 36, no. 8 (2007): 614–221; Johnny Gasperi et al., "Microplastics in Air: Are We Breathing It In?," *Current Opinion in Environmental Science & Health* 1 (2018): 1–5; Albert A. Koelmans et al., "Microplastics in Freshwaters and Drinking Water: Critical Review and Assessment of Data Quality," *Water Research* 155 (2019): 410–422.

7. Timothy Morton, ed., *Cultures of Taste / Theories of Appetite: Eating Romanticism* (New York: Palgrave Macmillan, 2004), 1.

8. Klervia Jaouen et al., "Exceptionally High δ15N Values in Collagen Single Amino Acids Confirm Neandertals as High-Trophic Level Carnivores," *Proceedings of the National Academy of Sciences* 116, no. 11 (2019): 4928–4933.

9. DK, *The Story of Food: An Illustrated History of Everything We Eat*, foreword by Giles Coren (New York: DK Publishing, 2018), 9.

10. Dorothée G. Drucker et al., "Isotopic Analyses Suggest Mammoth and Plant in the Diet of the Oldest Anatomically Modern Humans from Far Southeast Europe," *Scientific Reports* 7, no. 1 (2017): 1–10.

11. Bennett et al., "The Broiler Chicken," 2.

12. M. Shahbandeh, "Poultry: Number of Chickens Worldwide, 2018," Statista, February 5, 2021, https://www.statista.com/statistics/263962/number-of-chickens-worldwide-since-1990/.

13. Bennett et al., "The Broiler Chicken," 7.

14. H. L. Shrader, "The Chicken-of-Tomorrow Program: Its Influence on 'Meat-Type' Poultry Production," *Poultry Science* 31, no. 1 (1952): 3–10.

15. Bennett et al., "The Broiler Chicken," 7.

16. Bennett et al., "The Broiler Chicken," 7.

17. Hannah Ritchie and Max Roser, "Meat and Dairy Production," Our World in Data, last modified November 2019, https://ourworldindata.org/meat-production.

18. Bennett et al., "The Broiler Chicken," 8.

19. Koen Vanmechelen, *Cosmopolitan Chicken Project (CCP)*, 1999, accessed August 13, 2021, https://www.koenvanmechelen.be/cosmopolitan-chicken-project-ccp.

20. Vanmechelen, *Cosmopolitan Chicken Project*.

21. Nonhuman Nonsense, *Pink Chicken Project*, 2017, accessed August 25, 2021, https://pinkchickenproject.com/.

22. Anthony D. Barnosky et al., "Has the Earth's Sixth Mass Extinction Already Arrived?," *Nature* 471, no. 7336 (2011): 51.

23. Gerardo Ceballos, Paul R. Ehrlich, and Rodolfo Dirzo, "Biological Annihilation via the Ongoing Sixth Mass Extinction Signaled by Vertebrate Population Losses and Declines," *PNAS* 114, no. 30 (2017): 6094.

24. David Nogués-Bravo et al., "Climate Change, Humans, and the Extinction of the Woolly Mammoth," *PLoS Biology* 6, no. 4 (2008): 685–692.

25. S. T. Turvey and C. L. Risley, "Modelling the Extinction of Steller's Sea Cow," *Biology Letters* 2, no. 1 (2006): 94–97.

26. Turvey and Risley, "Modelling the Extinction of Steller's Sea Cow."

27. Lenore Newman, *Lost Feast: Culinary Extinction and the Future of Food* (Toronto: ECW Press, 2019).

28. Newman, *Lost Feast*.

29. Vanda Felbab-Brown, *The Extinction Market: Wildlife Trafficking and How to Counter It* (Oxford: Oxford University Press, 2017), 221–225.

30. Andrea Dell'Apa, M. Chad Smith, and Mahealani Y. Kaneshiro-Pineiro, "The Influence of Culture on the International Management of Shark Finning," *Environmental Management* 54, no. 2 (2014): 151–161.

31. Alexander C. Lees and Stuart L. Pimm, "Species, Extinct before We Know Them?," *Current Biology* 25, no. 5 (2015): 177–180.

32. Miriam Songster, *GhostFood*, 2019, Accessed August 3, 2021. https://songster.net/projects/ghostfood/.

33. Jeffrey A. Hutchings and John D. Reynolds, "Marine Fish Population Collapses: Consequences for Recovery and Extinction Risk," *BioScience* 54, no. 4 (2004): 297.

34. Anthony Ricciardi, "Invasive Species," in *Encyclopedia of Sustainability Science and Technology*, ed. Robert A. Meyers (New York: Springer, 2012), 162.

35. Martin A. Nuñez et al., "Invasive Species: To Eat or Not to Eat, That Is the Question," *Conservation Letters* 5, no. 5 (2012): 334–341.

36. Nuñez et al., "Invasive Species," 335.

37. Nuñez et al., "Invasive Species," 334–335.

38. Nuñez et al., "Invasive Species," 337.

39. Nuñez et al., "Invasive Species," 337.

40. Timothy Morton, *The Ecological Thought* (Cambridge, MA: Harvard University Press, 2012), 130–135.

41. C. N. Waters et al., "The Anthropocene Is Functionally and Stratigraphically Distinct from the Holocene," *Science* 351, no. 6269 (2016): 137–147; Jan Zalasiewicz et al., "When Did the Anthropocene Begin? A Mid-Twentieth Century Boundary Level Is Stratigraphically Optimal," *Quaternary International* 383 (2015): 196–203.

42. Barbara Hoeling, Douglas Reed, and P. B. Siegel, "Going Bananas in the Radiation Laboratory," *American Journal of Physics* 67, no. 5 (1999): 440.

43. Maristela Martins et al., "Brazil Nuts: Determination of Natural Elements and Aflatoxin," *Acta Amazonica* 42, no. 1 (2012): 157–164.

44. Waters et al., "The Anthropocene," 142; Zalasiewicz et al., "When Did the Anthropocene Begin?," 5–6.

45. D. J. Madigan, Z. Baumann, and N. S. Fisher, "Pacific Bluefin Tuna Transport Fukushima-Derived Radionuclides from Japan to California," *Proceedings of the National Academy of Sciences* 109, no. 24 (2012): 9483–9486.

46. Michaela Čadová et al., "Radioactivity in Mushrooms from Selected Locations in the Bohemian Forest, Czech Republic," *Radiation and Environmental Biophysics* 56, no. 2 (2017): 167–175.

47. Ph. Hubert et al., "Radioactivity Measurements Applied to the Dating and Authentication of Old Wines," *Comptes Rendus Physique* 10, no. 7 (2009): 622–629.

48. Petra Kozjak and Vladimir Meglič, "Mutagenesis in Plant Breeding for Disease and Pest Resistance," in *Mutagenesis*, ed. Rajnikant Mishra (Rijeka, Croatia: InTech, 2012), 195; U. Lundqvist, "Eighty Years of Scandinavian Barley Mutation Genetics and Breeding," in *Induced Plant Mutations in the Genomics Era*, ed. Q. Y. Shu (Rome: FAO, 2009), 39.

49. Özge Çelik and Çimen Atak, "Applications of Ionizing Radiation in Mutation Breeding," in *New Insights on Gamma Rays*, ed. Ahmed M. Maghraby (London: InTech, 2017), 115.

50. Filippo Tommaso Marinetti, *The Futurist Cookbook*, ed. Lesley Chamberlain, trans. Suzanne Brill (London: Penguin Books, 2014; first published in Italian in 1932 as *La Cucina Futurista*), 39 (citations refer to the Penguin edition).

51. The Center for Genomic Gastronomy, *Cobalt 60 Sauce*, 2013, accessed August 3, 2021, https://genomicgastronomy.com/work/2013-2/cobalt-60-sauce.

52. Timothy Morton, *Hyperobjects: Philosophy and Ecology after the End of the World* (Minneapolis: University of Minnesota Press, 2017), 1–3, 125.

212   Notes

53. Jan Zalasiewicz et al., "The Geological Cycle of Plastics and Their Use as a Stratigraphic Indicator of the Anthropocene," *Anthropocene* 13 (2016): 3.

54. Laura Parker, "A Whopping 91% of Plastic Isn't Recycled," *National Geographic*, December 20, 2018.

55. Patricia L. Corcoran, Charles J. Moore, and Kelly Jazvac, "An Anthropogenic Marker Horizon in the Future Rock Record," *GSA Today* 24, no. 6 (2014): 4–8.

56. Koelmans et al., "Microplastics in Freshwaters," 411.

57. Zalasiewicz et al., "The Geological Cycle of Plastics," 5.

58. Manuel Arias-Maldonado, "The Anthropocenic Turn: Theorizing Sustainability in a Postnatural Age," *Sustainability* 8, no. 1 (2015): 10.

59. Joe Yates et al., "PROTOCOL: Plastics in the Food System: Human Health, Economic and Environmental Impacts. A Scoping Review," *Campbell Systematic Reviews* 15, no. 1–2 (2019): e1033.

60. World Economic Forum, "The New Plastics Economy: Rethinking the Future of Plastics," World Economic Forum, 2016, 7.

61. Sarah Knapton, "Fish Eat Plastic in the Ocean Because It Smells like Food, Scientists Discover," *The Telegraph*, August 16, 2017; Matthew S. Savoca et al., "Odours from Marine Plastic Debris Induce Food Search Behaviours in a Forage Fish," *Proceedings of the Royal Society B: Biological Sciences* 284, no. 1860 (2017): 1–8.

62. David M. Boje, *Storytelling in the Global Age: There Is No Planet B* (Hackensack, NJ: World Scientific, 2019), ix.

63. Ali Karami et al., "The Presence of Microplastics in Commercial Salts from Different Countries," *Scientific Reports* 7, no. 1 (2017): 46173.

64. Gerd Liebezeit and Elisabeth Liebezeit, "Non-Pollen Particulates in Honey and Sugar," *Food Additives & Contaminants: Part A* 30, no. 12 (2014): 2136–2140; Peter Mühlschlegel et al., "Lack of Evidence for Microplastic Contamination in Honey," *Food Additives & Contaminants: Part A* 34, no. 11 (2017): 1982–1989.

65. Gerd Liebezeit and Elisabeth Liebezeit, "Synthetic Particles as Contaminants in German Beers," *Food Additives & Contaminants: Part A* 31, no. 9 (2014): 1574–1578; Mary Kosuth, Sherri A. Mason, and Elizabeth V. Wattenberg, "Anthropogenic Contamination of Tap Water, Beer, and Sea Salt," *PLoS One* 13, no. 4 (2018): e0194970.

66. Gasperi et al., "Microplastics in Air."

67. Koelmans et al., "Microplastics in Freshwaters."

68. Madeleine Smith et al., "Microplastics in Seafood and the Implications for Human Health." *Current Environmental Health Reports* 5, no. 3 (2018): 375–386.

69. Shosuke Yoshida et al., "A Bacterium That Degrades and Assimilates Poly(Ethylene Terephthalate)," *Science* 351, no. 6278 (2016): 1196–1199.

70. Jun Yang et al., "Evidence of Polyethylene Biodegradation by Bacterial Strains from the Guts of Plastic-Eating Waxworms," *Environmental Science & Technology* 48, no. 23 (2014): 13776–13784.

71. Yoshida et al., "A Bacterium That Degrades."

72. LIVIN Studio, *Fungi Mutarium*, 2014, accessed August 3, 2021, http://www.livinstudio.com/fungi-mutarium.

73. The Center for Genomic Gastronomy, *Smog Tasting*, 2011, accessed August 10, 2021, https://genomicgastronomy.com/work/2011-2/smog-tasting.

74. Marielle Dubbeling and Henk de Zeeuw, "Urban Agriculture and Climate Change Adaptation: Ensuring Food Security through Adaptation," in *Resilient Cities: Cities and Adaptation to Climate Change—Proceedings of the Global Forum 2010*, ed. Konrad Otto-Zimmermann (Dordrecht: Springer, 2011), 441–450.

75. Alexander D. Waffle et al., "Urban Heat Islands as Agricultural Opportunities: An Innovative Approach," *Landscape and Urban Planning* 161 (2017): 103–114.

76. Waffle et al., "Urban Heat Islands."

77. Jan Cohrs and Morgan Levy, *Alviso's Medicinal All-Salt: by Jon Cohrs and Morgan Levy*, 2010, accessed August 3, 2021, http://all-salt.splnlss.com.

Chapter 4

1. Elaine Showalter, *Hystories: Hysterical Epidemics and Modern Culture* (London: Picador, 1997), 3.

2. Maria Amalia Munoz-Pineiroa and Brigitte Toussaint, "'Fake Rice' on African and Asian Markets: Rumour or Evidence? Factsheet—2017," European Commission, 2018, JCR110625; Ataul Goni Rabbani et al., "Investigating the Existence of Artificial Eggs in Bangladesh and the Fact," *Journal of Applied Sciences* 19, no. 7 (2019): 701–707.

3. S. M. Zahid Hosen, Swati Paul, and Dibyajyoti Saha, "Artificial and Fake Eggs: Dance of Death," *Advances in Pharmacology and Pharmacy* 1, no. 1 (2013): 13–17.

4. Fred Gale and Jean C. Buzby, "Imports from China and Food Safety Issues," *United States Department of Agriculture, Economic Research Service, Economic Information Bulletin*, July 2009, 2.

5. Munoz-Pineiroa and Toussaint, "'Fake Rice' on African and AsianMarkets."

6. Rabbani et al., "Investigating the Existence of Artificial Eggs."

7. Hosen, Paul, and Saha, "Artificial and Fake Eggs," 16.

8. Rabbani et al., "Investigating the Existence of Artificial Eggs."

9. Nate Archer, *Matt Brown: Future of Food*, Designboom, accessed August 17, 2021, https://www.designboom.com/design/matt-brown-future-of-food/.

10. Sanche de Gramont, "Popular Cheese in Italy Exposed as Garbage," *Washington Post*, September 20, 1962.

11. De Gramont, "Popular Cheese in Italy."

12. Michael H. Tunick, *The Science of Cheese* (New York: Oxford University Press, 2014), 202.

13. Lydia Mulvany, "The Parmesan Cheese You Sprinkle on Your Penne Could Be Wood," Bloomberg, February 16, 2016, https://www.bloomberg.com/news/articles/2016-02-16/the-parmesan-cheese-you-sprinkle-on-your-penne-could-be-wood.

14. Anqi Shen, "'Being Affluent, One Drinks Wine': Wine Counterfeiting in Mainland China," *International Journal for Crime, Justice and Social Democracy* 7, no. 4 (2018): 16–32.

15. Shen, "'Being Affluent, One Drinks Wine,'" 21.

16.   Shen, "'Being Affluent, One Drinks Wine,'" 20.

17.   Science Communication Unit, *Science for Environment Policy In-Depth Report: Sustainable Food*, Report produced for the European Commission DG Environment (Bristol: University of the West of England, 2013), 3.

18.   Mario Herrero et al., "Livestock and the Environment: What Have We Learned in the Past Decade?," *Annual Review of Environment and Resources* 40, no. 1 (2015): 179.

19.   Herrero et al., "Livestock and the Environment," 180.

20.   David Tilman et al., "Global Food Demand and the Sustainable Intensification of Agriculture," *PNAS* 108, no. 50 (2011): 20260.

21.   H. C. J. Godfray and Oxford Martin School, Oxford University, "Meat: The Future Series— Alternative Proteins," World Economic Forum, 2019, 6, https://eprints.whiterose.ac.uk/170474.

22.   Godfray and Oxford Martin School, "Alternative Proteins," 17.

23.   Godfray and Oxford Martin School, "Alternative Proteins," 10.

24.   European Parliament and Council of the European Union, "Regulation (EU) No 1308/2013: Establishing a Common Organisation of the Markets in Agricultural Products and Repealing Council Regulations (EEC) No 922/72, (EEC) No 234/79, (EC) No 1037/2001 and (EC) No 1234/2007," *Official Journal of the European Union* (December 2013): 814–815.

25.   European Parliament, "Are Veggie Burgers or Tofu Steaks Going to Be Banned?," News: European Parliament, June 28, 2021, https://www.europarl.europa.eu/news/en/press-room/202010 19BKG89682/eu-farm-policy-reform-as-approved-by-meps/7/are-veggie-burgers-or-tofu-steaks-going-to-be-banned.

26.   Impossible Foods, "Impossible Foods Introduces Impossible Burger at Momofuku Nishi," press release, July 2016, https://assets.ctfassets.net/hhv516v5f7sj/44RJd5HZ2zciumW5WU6eoS/ 5f5a0f2c37470c5c7b3b60e6a0dc6e83/07_2016_IF_MOMOFUKU_NISHI_LAUNCH.pdf.

27.   Lenore Newman, *Lost Feast: Culinary Extinction and the Future of Food* (Toronto: ECW Press, 2019), chap. 2, EPUB.

28.   University of the Arts London, *Material Futures* (London: University of the Arts London, 2016), 45–46.

29.   Winston Churchill, "Fifty Years Hence," *Strand Magazine*, December 1931, 200.

30.   Russell Ross, "The Smooth Muscle Cell," *Journal of Cell Biology* 50, no. 1 (1971): 172.

31.   Oron Catts and Ionat Zurr, "Disembodied Livestock: The Promise of a Semi-Living Utopia," *Parallax* 19, no. 1 (2013): 101–113.

32.   Oron Catts and Ionat Zurr, *Disembodied Cuisine*, Tissue, Culture & Art Project, 2003, accessed August 13, 2021, https://tcaproject.net/portfolio/disembodied-cuisine/.

33.   David A. Brindley et al., "Peak Serum: Implications of Serum Supply for Cell Therapy Manufacturing," *Future Medicine* 7, no. 1 (2012): 10.

34.   Catts and Zurr, "Disembodied Livestock," 107.

35.   Femke Kools, "What's Been Going on with the 'Hamburger Professor,'" News & Events, Maastricht University, April 11, 2019, https://www.maastrichtuniversity.nl/news/what%E2%80% 99s-been-going-%E2%80%98hamburger-professor%E2%80%99.

36. Catts and Zurr, "Disembodied Livestock," 8.

37. Lori Andrews, "Tissue Culture: The Line between Art and Science Blurs When Two Artists Hang Cells in Galleries," *Journal of Life Sciences* (September 2007): 73.

38. Melissa Hogenboom, "What Does a Stem Cell Burger Taste Like?," BBC, August 5, 2013, https://www.bbc.com/news/science-environment-23529841.

39. Damian Carrington, "No-Kill, Lab-Grown Meat to Go on Sale for First Time," *The Guardian*, December 2, 2020.

40. Kristopher Gasteratos, "Nature & the Neomnivore," *The Cellular Agriculture Environmental Impact Compendium*, 2017, 67.

41. Liz Allen, "Lab-Grown Tuna: Freaky or Revolutionary? Either Way, It's Here," *Forbes*, April 10, 2020.

42. Fransiska Zakaria-Runkat, Wanchai Worawattanamateekul, and Ong-Ard Lawhavinit, "Production of Fish Serum Products as Substitute for Fetal Bovine Serum in Hybridoma Cell Cultures from Surimi Industrial Waste," *Kasetsart Journal: Natural Science* 40 (2006): 198–205.

43. Vanda Felbab-Brown, *The Extinction Market: Wildlife Trafficking and How to Counter It* (Oxford: Oxford University Press, 2017), 222; Andrea Dell'Apa, M. Chad Smith, and Mahealani Y. Kaneshiro-Pineiro, "The Influence of Culture on the International Management of Shark Finning," *Environmental Management* 54, no. 2 (2014): 151–161.

44. Kuang-Yi Ku, *Tiger Penis Project*, 2018, accessed August 14, 2021, https://www.kukuangyi.com/tiger-penis-project.html.

45. Arnold van Huis et al., *Edible Insects: Future Prospects for Food and Feed Security* (Rome: Food and Agriculture Organization of the United Nations, 2013), xiii.

46. Oron Catts, Ionat Zurr, and Robert Fosk, *Stir Fly: Nutrient Bug 1.0*, 2016, accessed September 1, 2021, https://tcaproject.net/portfolio/stir-fly-nutrient-bug-1-0.

47. DK, *The Story of Food: An Illustrated History of Everything We Eat*, foreword by Giles Coren (New York: DK Publishing, 2018), 9.

48. Jiangyong Lu and Zhigang Tao, "Sanlu's Melamine-Tainted Milk Crisis in China," Asia Case Research Centre, University of Hong Kong, 2009, 1–24.

49. Lu and Tao, "Sanlu's Melamine-Tainted Milk," 2.

50. Review of *Eat Not This Flesh: Food Avoidances from Prehistory to the Present* by Frederick J. Simoons, in *Whole Earth Catalog*, ed. Peter Warshall, Summer 1999, http://www.wholeearth.com/issue/2097/book-review/196/eat.not.this.flesh.food.avoidances.from.prehistory.to.the.present (site discontinued).

51. Arwa Mahdawi, "What If Eating Meat Is Not Only Wrong—but Obsolete?," *The Guardian*, June 5, 2019.

52. Koert van Mensvoort and Hendrik-Jan Grievink, *The In Vitro Meat Cookbook* (Amsterdam: BIS Publishers and Next Nature Network, 2014).

53. Open Meals, *Digital Oden*, 2018, accessed September 1, 2021, https://www.open-meals.com/digitaloden.

54. Open Meals, *Digital Oden*.

55. Azarmidokht Gholamipour-Shirazi et al., "How to Formulate for Structure and Texture via Medium of Additive Manufacturing: A Review," *Foods* 9, no. 4 (2020): 6.

56. Gholamipour-Shirazi et al., "How to Formulate for Structure," 9–13.

57. NASA Spinoff, "Deep-Space Food Science Research Improves 3D-Printing Capabilities," National Aeronautics and Space Administration (NASA), 2019, accessed January 5, 2021, https://spinoff.nasa.gov/Spinoff2019/ip_2.html.

58. Karna Ramachandraiah, "Potential Development of Sustainable 3D-Printed Meat Analogues: A Review," *Sustainability* 13, no. 2 (2021): 938.

59. Meydan Levy, *Neo Fruit Concept*, 2019, Meydanish, https://meydanish.wixsite.com/portpoliome/concept.

60. Levy, *Neo Fruit Concept*.

61. Bitelabs, *Eat Celebrity Meat*, accessed September 3, 2021, http://bitelabs.org.

Chapter 5

1. Ron Sender, Shai Fuchs, and Ron Milo, "Revised Estimates for the Number of Human and Bacteria Cells in the Body," *PLoS Biology* 14, no. 8 (2016), 1, 9.

2. L. Durso and R. Hutkins, "Starter Cultures," in *Encyclopedia of Food Sciences and Nutrition*, 2nd ed., ed. Benjamin Caballero (San Diego, CA: Academic Press, 2003), 5583–5593.

3. Christina Agapakis, *Selfmade*, 2013, accessed September 7, 2021, https://www.agapakis.com/work/selfmade.

4. Agapakis, *Selfmade*.

5. Agapakis, *Selfmade*.

6. Ruby Tandoh, "Ruby Tandoh: How I Was Turned into a Human Cheese," *The Guardian*, May 13, 2019.

7. Cristiane C. P. Andrade et al., "Microbial Dynamics during Cheese Production and Ripening: Physicochemical and Biological Factors," *Food Global Science Books* 2, no. 2 (2008): 93.

8. Andrade et al., "Microbial Dynamics during Cheese Production," 93.

9. David A. Relman, "New Technologies, Human-Microbe Interactions, and the Search for Previously Unrecognized Pathogens," *Journal of Infectious Diseases* 186, no. 2 (2002): 256.

10. Rosanne Yente Hertzberger, "Is It Really Possible to Make Yogurt from Vaginal Bacteria?," Rosanne Hertzberger (blog), accessed September 4, 2021, rosannehertzberger.nl/2015/03/31/is-it-really-possible-to-make-yogurt-from-vaginal-bacteria/.

11. Janet Jay, "How to Make Breakfast with Your Vagina," Vice, February 9, 2015, https://www.vice.com/en/article/mgbx9q/how-to-make-breakfast-with-your-vagina.

12. Hertzberger, "Is It Really Possible to Make Yogurt from Vaginal Bacteria?"

13. Jay, "How to Make Breakfast."

14. Toi Sennhauser, *Mama's Natural Breakfast*, accessed November 18, 2020, http://toisennhauser.com/work/13/mamas-natural-breakfast (site discontinued).

15. Megan Friedman, "This Woman Is Making Sourdough Bread Using Yeast from Her Vagina," *Cosmopolitan*, November 24, 2015.

16. Toi Sennhauser, *Mama's Natural Breakfast*.

17. Brendan Kiley, "From Vagina Beer to Socrates," The Stranger, October 13, 2005, https://www.the stranger.com/seattle/theater-news/Content?oid=23512.

18. Laure Ségurel and Céline Bon, "On the Evolution of Lactase Persistence in Humans," *Annual Review of Genomics and Human Genetics* 18, no. 1 (2017): 297–319.

19. Rainer Haas et al., "Cow Milk versus Plant-Based Milk Substitutes: A Comparison of Product Image and Motivational Structure of Consumption," *Sustainability* 11, no. 18 (2019): 1–25.

20. David Julian McClements, Emily Newman, and Isobelle Farrell McClements, "Plant-Based Milks: A Review of the Science Underpinning Their Design, Fabrication, and Performance," *Comprehensive Reviews in Food Science and Food Safety* 18, no. 6 (2019): 2049.

21. McClements, Newman, and McClements, "Plant-Based Milks," 2049–2051.

22. Katie Hinde and Lauren A. Milligan, "Primate Milk: Proximate Mechanisms and Ultimate Perspectives," *Evolutionary Anthropology: Issues, News, and Reviews* 20, no. 1 (2011): 10.

23. Mariana Meneses Romero, "Eating Human Cheese: The Lady Cheese Shop (Est. 2011)," *Feast Journal* no. 3 (2016).

24. Romero, "Eating Human Cheese."

25. Romero, "Eating Human Cheese."

26. Zack Denfeld and Cathrine Kramer, "Feeding Dangerous Ideas," interview by Zane Cerpina, *EE Experimental Emerging Art* no. 3 (2018): 48.

27. Mars El Brogy, "Breast Milk Ice Cream Is Back," *The Independent*, April 24, 2015.

28. Alison Lynch, "Someone's Re-Launched Breast Milk Ice Cream in Honour of the Royal Baby," Metro, May 5, 2015, https://metro.co.uk/2015/04/23/breastfeeding-campaigner-re-launches-breast -milk-ice-cream-to-celebrate-the-birth-of-the-royal-baby-5163591; "Matt O'Connor: Father for Justice and Ice Cream Extremist," *Evening Standard*, April 10, 2012.

29. O. Flidel-Rimon and E. S. Shinwell, "Breast Feeding Twins and High Multiples," *Archives of Disease in Childhood: Fetal and Neonatal Edition* 91, no. 5 (2006): 378.

30. Julian Roche, "Farm Management," in *Agribusiness: An International Perspective* (Abingdon, UK: Routledge, 2020), 140.

31. Bin Yang et al., "Characterization of Bioactive Recombinant Human Lysozyme Expressed in Milk of Cloned Transgenic Cattle," *PLoS ONE* 6, no. 3 (2011).

32. Costas N. Karatzas and Jeffrey D. Turner, "Toward Altering Milk Composition by Genetic Manipulation: Current Status and Challenges," *Journal of Dairy Science* 80, no. 9 (1997): 2229–2230.

33. Mathilde Cohen, "The Lactating Man," in *Making Milk: The Past, Present, and Future of Our Primary Food*, ed. Mathilde Cohen and Yoriko Otomo (London: Bloomsbury Academic, 2017), 141–160.

34. Jerzy K. Kulski and Peter E. Hartmann, "Composition of Breast Fluid of a Man with Galactorrhea and Hyperprolactinaemia," *Journal of Clinical Endocrinology & Metabolism* 52, no. 3 (1981): 581–582.

35. Jesse J. Edwards, Sandra L. Tollaksen, and Norman G. Anderson, "Proteins of Human Semen. I. Two-Dimensional Mapping of Human Seminal Fluid," *Clinical Chemistry* 27, no. 8 (1981): 1335–1340.

36. Derek H. Owen and David F. Katz, "A Review of the Physical and Chemical Properties of Human Semen and the Formulation of a Semen Simulant," *Journal of Andrology* 26, no. 4 (2005): 461.

37. Paul "Fotie" Photenhauer, *Natural Harvest: A Collection of Semen-Based Recipes* (North Charleston, SC: Createspace Independent Publishing Platform, 2008).

38. Owen and Katz, "A Review of the Physical and Chemical Properties," 465.

39. Owen and Katz, "A Review of the Physical and Chemical Properties," 462.

40. Vincenzo Savica et al., "Urine Therapy through the Centuries," *Journal of Nephrology* 24, no. Suppl. 17 (2011): 123.

41. United States Department of the Army, *US Army Field Manual 3-05.70: Survival* (Washington, DC: Government Printing Office, 2002), 64.

42. National Aeronautics and Space Administration (NASA), "NASA Gives Space Station Crew 'Go' to Drink Recycled Water," press release 09-096, NASA, May 20, 2009, https://www.nasa.gov/home/hqnews/2009/may/HQ_09-096_Recycled_Water_Go.html.

43. C. Rose et al., "The Characterization of Feces and Urine: A Review of the Literature to Inform Advanced Treatment Technology," *Critical Reviews in Environmental Science and Technology* 45, no. 17 (2015): 1850.

44. Enid Contes, "Water Recycling," National Aeronautics and Space Administration (NASA), October 17, 2014, https://www.nasa.gov/content/water-recycling; CSA ASC, *Astronauts Drink Urine and Other Waste Water*, YouTube, 2013, https://youtu.be/ZQ2T9OJY1lg.

45. John Schwartz, "Tasting NASA's Recycled Water," The Lede (blog), *New York Times*, November 13, 2008.

46. Schwartz, "Tasting NASA's Recycled Water."

47. Mariya Lobanovska and Giulia Pilla, "Penicillin's Discovery and Antibiotic Resistance: Lessons for the Future?," *Yale Journal of Biology and Medicine* 90, no. 1 (2017): 136.

48. Régine Debatty, "Family Whisky," We Make Money Not Art (blog), August 23, 2010, https://we-make-money-not-art.com/gilpin_family_whisky.

49. Rose et al., "The Characterization of Feces and Urine," 1832.

50. Rose et al., "The Characterization of Feces and Urine," 1840.

51. Rose et al., "The Characterization of Feces and Urine," 1840.

52. Lisa M. Steinberg, Rachel E. Kronyak, and Christopher H. House, "Coupling of Anaerobic Waste Treatment to Produce Protein- and Lipid-Rich Bacterial Biomass," *Life Sciences in Space Research* 15 (2017): 32–42.

53. National Aeronaustics and Space Administration (NASA), "NASA Awards Grants for Technologies That Could Transform Space Exploration," press release 15-170, NASA, August 14, 2015.

54. Liam Jackson, "Microbes May Help Astronauts Transform Human Waste into Food," Penn State News, January 25, 2018, https://news.psu.edu/story/502406/2018/01/25/research/microbes-may-help-astronauts-transform-human-waste-food.

55.   Nadia Arumugam, "Meat Made from Human Feces: Hoax or Japan's Best New Invention?," *Forbes*, August 9, 2011.

56.   Daily Mail Reporter, "Japanese Scientist Makes a 'Delicious' Burger out of . . . Human Excrement," *Daily Mail*, June 17, 2011, https://www.dailymail.co.uk/sciencetech/article-2004791/Japanese-scientist-makes-delicious-burger--human-EXCREMENT.html.

57.   Lina Zeldovich, "A History of Human Waste as Fertilizer," JSTOR Daily, November 18, 2019, https://daily.jstor.org/a-history-of-human-waste-as-fertilizer.

58.   WHO, "Obesity and Overweight," June 9, 2021, accessed September 6, 2021, https://www.who.int/news-room/fact-sheets/detail/obesity-and-overweight.

59.   International Society of Aesthetic Plastic Surgery (ISAPS), "ISAPS International Survey on Aesthetic/Cosmetic Procedures Performed in 2019," 2020, 6.

60.   ISAPS, "ISAPS International Survey," 6.

61.   Warwick L. Greville, "Proposed Safety Guidelines for the Maximum Volume Fat Removal by Tumescent Liposuction," in *Liposuction*, ed. Melvin A. Shiffman and Alberto Di Giuseppe (Berlin: Springer, 2006), 30.

62.   Zoran Todorović, *Assimilation*, 2017, https://www.zorantodorovic.com/portfolio_page/assimilation/.

63.   Zoran Todorović, "Human Gourmet," interview by Zane Cerpina, *EE Experimental Emerging Art* (2018): 50–51.

64.   Evaristti Studios, *Polpette al Grasso di Marco*, accessed September 28, 2020, https://www.evaristti.com/polpette-al-grasso-di-marco.

65.   Marko Marković, *Selfeater / The Thirst*, 2012, accessed September 2, 2021, http://markovichmarko.blogspot.com/2012/02/selfeater-thirst-production-2009-dopust.html.

66.   Richard Fleischer, dir., *Soylent Green*, 1973.

67.   Magnus Söderlund, "High-Level Climate Madness: About Eating Eating Human Flesh to Save the Climate," YouTube, 2019, https://www.youtube.com/watch?v=M-xJopfyJT8&ab_channel=ActionTW.

68.   James Cole, "Assessing the Calorific Significance of Episodes of Human Cannibalism in the Palaeolithic," *Scientific Reports* 7, no. 1 (2017): 3.

69.   Cole, "Assessing the Calorific Significance," 5.

70.   Adam Hadhazy, "Eat the Old: Could Mass Cannibalism Solve a Future Food Shortage?," Live Science, October 28, 2011, https://www.livescience.com/16779-soylent-green-real-life-cannibalism.html.

71.   Carys E. Bennett et al., "The Broiler Chicken as a Signal of a Human Reconfigured Biosphere," *Royal Society Open Science* 5, no. 12 (2018): 180325, 7.

72.   Matthew L. Richardson et al., "Causes and Consequences of Cannibalism in Noncarnivorous Insects," *Annual Review of Entomology* 55, no. 1 (2010): 39–53.

73.   Volker H. W. Rudolf and Janis Antonovics, "Disease Transmission by Cannibalism: Rare Event or Common Occurrence?," *Proceedings of the Royal Society B: Biological Sciences* 274, no. 1614 (2007): 1205; Simon Mead et al., "Balancing Selection at the Prion Protein Gene Consistent with Prehistoric Kurulike Epidemics," *Science* 300, no. 5619 (2003): 640–643.

74. Mead et al., "Balancing Selection at the Prion Protein Gene," 1, 3.

75. Rudolf and Antonovics, "Disease Transmission by Cannibalism," 1209.

76. Rich Wordsworth, "What's Wrong with Eating People?," Wired, October 28, 2017, https://www.wired.co.uk/article/lab-grown-human-meat-cannibalism.

77. Wordsworth, "What's Wrong with Eating People?"

78. Oron Catts, "The Art of the Semi-Living," January 2004, 1–2, https://www.researchgate.net/publication/242681743_The_Art_of_The_Semi-Living.

79. Theresa Schubert, *mEat Me*, 2020, accessed September 9, 2021, https://www.theresaschubert.com/works/meat-me.

80. BiteLabs, "BiteLabs," home page, accessed September 28, 2020, http://bitelabs.org.

81. The Center for Genomic Gastronomy, *To Flavour Our Tears*, 2016, accessed September 9, 2020, https://genomicgastronomy.com/work/2016-2/to-flavour-our-tears.

82. Center for Genomic Gastronomy, *To Flavour Our Tears*.

83. Center for Genomic Gastronomy, *To Flavour Our Tears*.

84. Center for Genomic Gastronomy, *To Flavour Our Tears*.

85. Center for Genomic Gastronomy, *To Flavour Our Tears*.

Chapter 6

1. Arnold Van Huis, "Edible Insects Contributing to Food Security?," *Agriculture & Food Security* 4, no. 1 (2015): 5.

2. H. C. J. Godfray et al., "Meat Consumption, Health, and the Environment," *Science* 361, no. 6399 (2018): 1.

3. Nikos Alexandratos and Jelle Bruinsma, "World Agriculture towards 2030/2050: The 2012 Revision," ESA Working Papers 288998, Food and Agriculture Organization of the United Nations, Agricultural Development Economics Division (2012): 77, 96.

4. Alexandratos and Bruinsma, "World Agriculture," 95.

5. Maeve Henchion et al., "Future Protein Supply and Demand: Strategies and Factors Influencing a Sustainable Equilibrium," *Foods* 6, no. 7 (2017): 2; Godfray et al., "Meat Consumption," 1.

6. Arnold Van Huis, "Potential of Insects as Food and Feed in Assuring Food Security," *Annual Review of Entomology* 58, no. 1 (2013): 575.

7. Lenka Kouřimská and Anna Adámková, "Nutritional and Sensory Quality of Edible Insects," *NFS Journal* 4 (2016): 22–23; Anna Rettore, Róisín M. Burke, and Catherine Barry-Ryan, "Insects: A Protein Revolution for the Western Human Diet," in *Dublin Gastronomy Symposium May 31–June 1, 2016*, 2016, 1; Arnold Van Huis, "Did Early Humans Consume Insects?," *Journal of Insects as Food and Feed* 3, no. 3 (2017): 162.

8. Arnold Van Huis et al., *Edible Insects: Future Prospects for Food and Feed Security* (Rome: Food and Agriculture Organization of the United Nations, 2013), 35–37.

9. Van Huis et al., *Edible Insects*, 15–33.

10.   Van Huis et al., *Edible Insects*, 59.

11.   Dennis G. A. B. Oonincx and Imke J. M. de Boer, "Environmental Impact of the Production of Mealworms as a Protein Source for Humans: A Life Cycle Assessment," *PLoS ONE* 7, no. 12 (2012): 4.

12.   Pier Miglietta et al., "Mealworms for Food: A Water Footprint Perspective," *Water* 7, no. 11 (2015): 6197.

13.   Van Huis, "Potential of Insects," 565; Lorenzo A. Cadinu et al., "Insect Rearing: Potential, Challenges, and Circularity," *Sustainability* 12, no. 11 (2020): 6, 4567.

14.   Cadinu et al., "Insect Rearing," 6.

15.   Attila Gere et al., "Insect Based Foods: A Nutritional Point of View," *Nutrition & Food Science International Journal* 4, no. 2 (2017): 2, 555638.

16.   Gere et al., "Insect Based Foods," 2.

17.   Rettore, Burke, and Barry-Ryan, "Insects: A Protein Revolution," 2.

18.   Leviticus 11:20–30.

19.   Van Huis et al., *Edible Insects*, 141–142.

20.   Van Huis et al., *Edible Insects*, xiii, 39, 126.

21.   Van Huis et al., *Edible Insects*, xiii, 39.

22.   Rettore, Burke, and Barry-Ryan, "Insects: A Protein Revolution," 2.

23.   Van Huis, "Potential of Insects," 573.

24.   F. V. Dunkel and C. Payne, "Introduction to Edible Insects," in *Insects as Sustainable Food Ingredients: Production, Processing and Food Applications*, ed. Aaron T. Dossey, Juan A. Morales-Ramos, and M. Guadalupe Rojas (Amsterdam: Academic Press, 2016), 12–13.

25.   Dunkel and Payne, "Introduction to Edible Insects," 17.

26.   Saara Maria Kauppi, Ida Nilstad Pettersen, and Casper Boks, "Consumer Acceptance of Edible Insects and Design Interventions as Adoption Strategy," *International Journal of Food Design* 4, no. 1 (2019): 39–62.

27.   Kauppi, Pettersen, and Boks, "Consumer Acceptance of Edible Insects," 39, 41.

28.   Saara-Maria Kauppi, "Packaging Design Strategies for Introducing Whole Mealworms as Human Food," in *DS-101: Proceedings of NordDesign 2020, Lyngby, Denmark, 12th–14th August 2020—Balancing Innovation and Operation*, ed. N. H. Mortensen, C. T. Hansen, and M. Deininger, Design Society, 2020.

29.   Kauppi, Pettersen, and Boks, "Consumer Acceptance of Edible Insects," 55; Rettore, Burke, and Barry-Ryan, "Insects: A Protein Revolution," 5.

30.   Kauppi, Pettersen, and Boks, "Consumer Acceptance of Edible Insects," 55.

31.   Sexy Food, *Concept*, accessed March 22, 2018, http://www.sexyfood.fr/en/ (site discontinued).

32.   Saara Maria Kauppi, "Insect Economy and Marketing: How Much and in What Way Could Insects Be Shown in Packaging?," master's thesis, Aalto University, 2016, 73.

33. Christina Hartmann et al., "The Psychology of Eating Insects: A Cross-Cultural Comparison between Germany and China," *Food Quality and Preference* 44 (2015): 148–156.

34. Susana Soares, *Insects Au Gratin*, accessed September 4, 2021, http://www.susanasoares.com/index.php?id=79.

35. Susana Soares and Andrew Forkes, "*Insects Au Gratin*: An Investigation into the Experiences of Developing a 3D Printer That Uses Insect Protein Based Flour as a Building Medium for the Production of Sustainable Food," in *DS 78: Proceedings of the 16th International Conference on Engineering and Product Design Education (E&PDE14), Design Education and Human Technology Relations, University of Twente, the Netherlands, September 4–5, 2014*, ed. Erik Bohemia, Arthur Eger, Wouter Eggink, Ahmed Kovacevic, Brian Parkinson, and Wessel Wits, Design Society, 2014, 426–431.

36. Harry McDade and C. Matilda Collins, "How We Might Overcome 'Western' Resistance to Eating Insects," in *Edible Insects*, ed. Heimo Mikkola (London: IntechOpen, 2020), 3.

37. Van Huis et al., *Edible Insects,* 1.

38. Julieta Ramos-Elorduy, *Creepy Crawly Cuisine: The Gourmet Guide to Edible Insects*, trans. Nancy Esteban (Rochester, VT: Park Street Press, 1998), 14–16.

39. Kouřimská and Adámková, "Nutritional and Sensory Quality of Edible Insects," 25.

40. Ramos-Elorduy, *Creepy Crawly Cuisine*, 14–16.

41. Ramos-Elorduy, *Creepy Crawly Cuisine*, 16.

42. Josh Evans et al., *On Eating Insects: Essays, Stories and Recipes* (London: Phaidon Press Limited, 2017).

43. Evans et al., *On Eating Insects*, 312–317.

44. Evans et al., *On Eating Insects,* 84, 222; Nordic Food Lab and the Cambridge Distillery, *Anty Gin*, 2013, October 11, 2014, https://nordicfoodlab.org/blog/2014/10/anty-gin.

45. Nordic Food Lab and the Cambridge Distillery, *Anty Gin*.

46. Van Huis et al., *Edible Insects*, 7.

47. Van Huis et al., *Edible Insects*, 36–37; Rettore, Burke, and Barry-Ryan, "Insects: A Protein Revolution, 2."

48. Van Huis et al., *Edible Insects*, 17.

49. Hiromu Ito et al., "Evolution of Periodicity in Periodical Cicadas," *Scientific Reports* 5, no. 1 (2015): 1–2, 14094.

50. John R. Cooley et al., "At the Limits: Habitat Suitability Modelling of Northern 17-Year Periodical Cicada Extinctions (*Hemiptera: Magicicada Spp.*)," *Global Ecology and Biogeography* 22, no. 4 (2013): 419.

51. Andrew M. Liebhold, Michael J. Bohne, and Rebecca L. Lilja, "Active Periodical Cicada Broods of the United States," USDA Forest Service Northern Research Station, Northeastern Area State & Private Forestry, 2013.

52. Cooley et al., "At the Limits," 411.

53. Lenore Newman, *Lost Feast: Culinary Extinction and the Future of Food* (Toronto: ECW Press, 2019).

54. Francisco Sánchez-Bayo and Kris A. G. Wyckhuys, "Worldwide Decline of the Entomofauna: A Review of Its Drivers," *Biological Conservation* 232 (2019): 8–27.

55. Sánchez-Bayo and Wyckhuys, "Worldwide Decline of the Entomofauna."

56. Andrew Paul Gutierrez and Luigi Ponti, "Analysis of Invasive Insects: Links to Climate Change," in *Invasive Species and Global Climate Change*, ed. Lewis H. Ziska and Jeffrey S. Dukes (Wallingford, UK: CABI, 2014), 45–61.

57. Anthony Ricciardi, "Invasive Species," in *Encyclopedia of Sustainability Science and Technology*, ed. Robert A. Meyers (New York: Springer, 2012), 162.

58. Martin A. Nuñez et al., "Invasive Species: To Eat or Not to Eat, That Is the Question," *Conservation Letters* 5, no. 5 (2012): 334–341.

59. Gengping Zhu et al., "Assessing the Ecological Niche and Invasion Potential of the Asian Giant Hornet," *PNAS* 117, no. 40 (2020): 24646.

60. Zhu et al., "Assessing the Ecological Nicke," 24646.

61. Hannah Sparks, "Deadly 'Murder Hornets' Are Also a Crunchy Gourmet Snack," *New York Post*, May 5, 2020.

62. Ben Dooley, "In Japan, the 'Murder Hornet' Is Both a Lethal Threat and a Tasty Treat," *New York Times*, May 5, 2020.

63. Sparks, "Deadly 'Murder Hornets'"; Dooley, "In Japan, the 'Murder Hornet.'"

64. René Cerritos, "Insects as Food: An Ecological, Social and Economical Approach," *CAB Reviews: Perspectives in Agriculture Veterinary Science Nutrition and Natural Resources* 4, no. 27 (2009).

65. Van Huis et al., *Edible Insects*, 26, 55.

66. Cerritos, "Insects as Food," 5.

67. Food and Agriculture Organization of the United Nations (FAO), *Desert Locust Crisis: Appeal for Rapid Response and Anticipatory Action in the Greater Horn of Africa* (Rome: FAO, 2020), 3.

68. Van Huis, "Potential of Insects," 569.

69. Van Huis et al., *Edible Insects*, 26.

70. Nera Kuljanic and Samuel Gregory-Manning, "What If Insects Were on the Menu in Europe?," Scientific Foresight Unit, European Parliamentary Research Unit, July 2020.

71. Dunkel and Payne, "Introduction to Edible Insects," 2.

72. Cadinu et al., "Insect Rearing," 2, 4567.

73. Dunkel and Payne, "Introduction to Edible Insects," 11–12.

74. United Nations, Department of Economic and Social Affairs, Population Division, *World Urbanization Prospects: The 2018 Revision (ST/ESA/SER.A/420)* (New York: United Nations, 2019), 1.

75. LIVIN Farms, *The Hive Magazine*, 2017, 30–37.

76. Kauppi, Pettersen, Boks, "Consumer Acceptance of Edible Insects," 54.

77. LIVIN Farms, *The Hive Magazine*, 2017, 14–20; Katharina Unger, *Farm 432: Insect Breeding*, 2013, accessed September 7, 2021, https://katharinaunger.com/farm-432-insect-breeding.

78. Búi Bjarmar Aðalsteinsson, *The Fly Factory*, 2014, accessed September 4, 2021, https://cargocollective.com/bjarmar/the-Fly-Factory.

79. Vassileios Varelas, "Food Wastes as a Potential New Source for Edible Insect Mass Production for Food and Feed: A Review," *Fermentation* 5, no. 3 (2019): 1–19, 81.

80. Nismah Nukmal et al., "Effect of Styrofoam Waste Feeds on the Growth, Development and Fecundity of Mealworms (*Tenebrio molitor*)," *OnLine Journal of Biological Sciences* 18, no. 1 (2018): 24–28.

81. Marc Paulusma, *BoeteBurger Project*, 2019, accessed September 4, 2021, https://www.studiomarc.org/made-by-mealworms.

82. Paulusma, *BoeteBurger Project*.

83. United States Environmental Protection Agency (EPA), "Frequent Questions Regarding EPA's Facts and Figures about Materials, Waste and Recycling," EPA, last updated July 29, 2021, accessed September 5, 2021, https://www.epa.gov/facts-and-figures-about-materials-waste-and-recycling/frequent-questions-regarding-epas-facts-and.

84. Slavoj Žižek, "The Truth about Turkey," Occupy Big Food, November 10, 2011, https://occupybigfood.wordpress.com/tag/boycott-butterball.

85. Ole G. Mouritsen and Klavs Styrbæk, *Mouthfeel: How Texture Makes Taste*, trans. Mariela Johansen (New York: Columbia University Press, 2017), 304.

86. Kashmira Gander, "A Designer Has Created a Bug-Eating Kit to Save Humanity," *The Independent*, February 16, 2017.

87. Samuel Imathiu, "Benefits and Food Safety Concerns Associated with Consumption of Edible Insects," *NFS Journal* 18 (2020): 4–7, 8.

88. Geoffrey Taylor and Nanxi Wang, "Entomophagy and Allergies: A Study of the Prevalence of Entomophagy and Related Allergies in a Population Living in North-Eastern Thailand," *Bioscience Horizons: The International Journal of Student Research* 11 (2018): 4–6; Kouřimská and Adámková, "Nutritional and Sensory Quality of Edible Insects," 25.

89. Taylor and Wang, "Entomophagy and Allergies," 2.

90. Imathiu, "Benefits and Food Safety Concerns," 5.

91. Imathiu, "Benefits and Food Safety Concerns," 6–7.

92. Taylor and Wang, "Entomophagy and Allergies," 7.

93. Anna-Liisa Elorinne et al., "Insect Consumption Attitudes among Vegans, Non-Vegan Vegetarians, and Omnivores," *Nutrients* 11, no. 2 (2019): 12, 292.

94. Kouřimská and Adámková, "Nutritional and Sensory Quality of Edible Insects," 25.

95. Van Huis, "Potential of Insects," 574.

96. Kuljanic and Gregory-Manning, "What If Insects Were on the Menu in Europe?," 2.

97. Jesse Erens et al., "A Bug's Life: Large-Scale Insect Rearing in Relation to Animal Welfare," VENIK, Wageningen, 2012, 13, 15.

98. Erens et al., "A Bug's Life," 9.

99. Arnold van Huis and Dennis G. Oonincx, "The Environmental Sustainability of Insects as Food and Feed: A Review," *Agronomy for Sustainable Development* 37, no. 5 (2017): 6–7, 43; Cadinu et al., "Insect Rearing," 14, 4567.

100. Cadinu et al., "Insect Rearing," 14.

101. Cadinu et al., "Insect Rearing," 14.

102. Cadinu et al., "Insect Rearing," 17.

103. Evans et al., *On Eating Insects*; Corey Mintz, "Why Eating Insects Won't End World Hunger," *Globe and Mail*, February 4, 2018.

104. Van Huis et al., *Edible Insects*, 45–46.

105. Van Huis et al., *Edible Insects*, 161.

Chapter 7

1. Amit Baratz, "The Source of the Gods' Immortality in Archaic Greek Literature," *Scripta Classica Israelica* 34 (2015): 151.

2. Rosanna Lauriola, "The Greeks and the Utopia: An Overview through Ancient Greek Literature," *Revista Espaço Acadêmico* 9, no. 97 (2009): 115.

3. Mark Dugdale, "The VR Shopping Experience Service Will Enable Local Businesses to Connect Stores, Virtually, to New and Existing Customers," VRWorldTech, March 25, 2020, https://vrworldtech.com/2020/03/25/stay-at-home-and-shop-thanks-to-lifestyles-in-360/.

4. David Reid, "Domino's Delivers World's First Ever Pizza by Drone," CNBC, November 16, 2016, https://www.cnbc.com/2016/11/16/dominos-has-delivered-the-worlds-first-ever-pizza-by-drone-to-a-new-zealand-couple.html.

5. Gyorgy Scrinis, *Nutritionism: The Science and Politics of Dietary Advice* (New York: Columbia University Press, 2013), 25–27.

6. Scrinis, *Nutritionism*, 94.

7. Scrinis, *Nutritionism*, 27–29.

8. Jana Strahler et al., "Orthorexia Nervosa: A Behavioral Complex or a Psychological Condition?," *Journal of Behavioral Addictions* 7, no. 4 (2018): 1143.

9. Joost Oude Groeniger et al., "Does Social Distinction Contribute to Socioeconomic Inequalities in Diet: The Case of 'Superfoods' Consumption," *International Journal of Behavioral Nutrition and Physical Activity* 14, no. 1 (2017): 1–2, 40.

10. National Health Service, "The Research on Superfoods Is Exaggerated by the Media," in *Superfoods*, ed. Roman Espejo (Farmington Hills, MI: Greenhaven Press, 2016), 34.

11. Steven Bratman, "Orthorexia vs. Theories of Healthy Eating," *Eating and Weight Disorders* 22 (2017): 381–385.

12. Strahler et al., "Orthorexia Nervosa," 1143.

13. Anna Williford, Barbara Stay, and Debashish Bhattacharya, "Evolution of a Novel Function: Nutritive Milk in the Viviparous Cockroach, *Diploptera punctata*," *Evolution and Development* 6, no. 2 (2004): 67.

14. Kamal Niaz, Jonathan Spoor, and Elizabeta Zaplatic, "Highlight Report: *Diploptera functata* (Cockroach) Milk as Next Superfood," *EXCLI Journal* 17 (2018): 721–723.

15. Kastalia Medrano, "Everyone Calm Down, Cockroach Milk Isn't Taking Over Just Yet," Inverse, August 1, 2016, https://www.inverse.com/article/19066-cockroach-milk-what-is-it.

16. Medrano, "Everyone Calm Down."

17. Warren James Belasco, "Future Notes: The Meal-in-a-Pill," in *Food in the USA: A Reader*, ed. Carole M. Counihan (New York: Routledge, 2002), 60.

18. Anna Bowman Dodd, *The Republic of the Future: Or, Socialism a Reality* (New York: Cassell, 1887); Belasco, "Future Notes," 62.

19. Dodd, *Republic of the Future*, 31.

20. Henry J. W. Dam, "Foods in the Year 2000: Professor Berthelot's Theory That Chemistry Will Displace," *McClure's Magazine*, 1894, 304.

21. David Butler, dir., *Just Imagine*, Hollywood, CA: Fox Film, 1930, film.

22. Butler, *Just Imagine*, video clip, YouTube, https://youtu.be/lcmg6JH3PXo.

23. Jason Ong Xiang An, "Future Typologies Part 1 | Restaurants | Now Serving: Future Foods," 2019, 20–21.

24. Belasco, "Future Notes," 60.

25. Soylent, home page, accessed September 5, 2021, https://soylent.com.

26. S. M. Tieken et al., "Effects of Solid versus Liquid Meal-Replacement Products of Similar Energy Content on Hunger, Satiety, and Appetite-Regulating Hormones in Older Adults," *Hormone and Metabolic Research* 39, no. 5 (2007): 389–394.

27. Josefin Larsson, "Foods of the Future: Gastropolitics and Climate Change in the Anthropocene," master's thesis, Department of Arts and Cultural Sciences, Lund University, 2018, 11.

28. Richard Faulk, *The Next Big Thing: A History of the Boom-or-Bust Moments That Shaped the Modern World* (San Francisco: Zest Books, 2015), 128.

29. Andrew F. Smith, *Encyclopedia of Junk Food and Fast Food* (Westport, CT: Greenwood Press, 2006), 221.

30. Annika J. Kettenburg et al., "From Disagreements to Dialogue: Unpacking the Golden Rice Debate," *Sustainability Science* 13, no. 5 (2018): 1470.

31. Karabi Datta et al., "Genetic Stability Developed for β-Carotene Synthesis in BR29 Rice Line Using Dihaploid Homozygosity," *PLoS ONE* 9, no. 6 (2014), 1.

32. Chureeporn Chitchumroonchokchai et al., "Potential of Golden Potatoes to Improve Vitamin A and Vitamin E Status in Developing Countries," *PLoS One* 12, no. 11 (2017): e0187102.

33. Golden Rice Humanitarian Board, "Golden Rice Project," Golden Rice Project, accessed September 6, 2021, http://www.goldenrice.org.

34. Andreas Greiner, *Monument for the 308*, 2016, accessed September 2, 2021, http://www.andreasgreiner.com/works/monument-for-the-308.

35. Greiner, *Monument for the 308*.

36. Paul Gong, *The Cow of Tomorrow*, 2015, accessed September 5, 2021, https://www.paulgong.co.uk/The-Cow-of-Tomorrow.

37. Alois Pfenniger et al., "Performance Analysis of a Miniature Turbine Generator for Intracorporeal Energy Harvesting," *Artificial Organs* 38, no. 5 (2014): 68–81.

38. Gong, *The Cow of Tomorrow.*

39. F. Jung et al., "*Spirulina platensis*, a Super Food?," *Journal of Cellular Biotechnology* 5, no. 1 (2019): 44.

40. Jung et al., "*Spirulina platensis*," 43.

41. Jung et al., "*Spirulina platensis*," 50–51.

42. Michael Burton and Michiko Nitta, *Algaculture*, 2010, accessed September 9, 2021, https://www.burtonnitta.co.uk/Algaculture.html.

43. Ines Barkia, Nazamid Saar, and Schonna R. Manning, "Microalgae for High-Value Products towards Human Health and Nutrition," *Marine Drugs* 17, no. 5 (2019): 4, 304.

44. Barkia, Saar, and Manning, "Microalgae," 3.

45. Mitch Kanter and Ashley Desrosiers, "Personalized Wellness Past and Future: Will the Science and Technology Coevolve?," *Nutrition Today* 54, no. 4 (2019): 175.

46. Kanter and Desrosiers, "Personalized Wellness," 178.

47. Open Meals, *Sushi Singularity*, 2019, accessed September 2, 2021, https://www.open-meals.com/sushisingularity/index_e.html.

Chapter 8

1.   World Health Oganization (WHO), "Fact Sheets - Malnutrition," WHO, accessed September 5, 2021, https://www.who.int/news-room/fact-sheets/detail/malnutrition.

2.   Andrew F. Smith, *Encyclopedia of Junk Food and Fast Food* (Westport, CT: Greenwood Press, 2006), 4, 247.

3.   G. Shridhar et al., "Modern Diet and Its Impact on Human Health," *Journal of Nutrition and Food Science* 5, no. 6 (2015): 2, 430.

4.   Shridhar et al., "Modern Diet," 2.

5.   Paul A. S. Breslin, "An Evolutionary Perspective on Food and Human Taste," *Current Biology* 23, no. 9 (2013): 415–416.

6.   Daniel Nettle, Clare Andrews, and Melissa Bateson, "Food Insecurity as a Driver of Obesity in Humans: The Insurance Hypothesis," *Behavioral and Brain Science* 40 (2017): e105; Fernando Sérgio Zucoloto, "Evolution of the Human Feeding Behavior," *Psychology & Neuroscience* 4, no. 1 (2011): 135; David A. Wiss, Nicole Avena, and Pedro Rada, "Sugar Addiction: From Evolution to Revolution," *Frontiers in Psychiatry* 9 (2018): 2, 545.

7.   Jana Strahler et al., "Orthorexia Nervosa: A Behavioral Complex or a Psychological Condition?," *Journal of Behavioral Addictions* 7, no. 4 (2018): 1144.

8.   Breslin, "An Evolutionary Perspective," 416.

9.   World Bank Group, *An Overview of Links between Obesity and Food Systems* (Washington, DC: World Bank, 2017), 4.

10.   Breslin, "An Evolutionary Perspective," 416.

11.   Breslin, "An Evolutionary Perspective," 11.

12.   Anna Thibodeaux, "16 Junk-Food Addicted Raccoons Are Relocated," *St. Charles Herald-Guide*, January 13, 2017.

13.   Agence France-Presse, "'We Live in a Cage': Residents Hide as Macaque 'Gangs' Take Over Thai City," *The Guardian*, June 24, 2020.

14.   Bernhard Warner, "Boar Wars: How Wild Hogs Are Trashing European Cities," *The Guardian*, July 30, 2019.

15.   Organisation for Economic Co-operation and Development (OECD), *The Heavy Burden of Obesity: The Economics of Prevention* (Paris: OECD, 2019), 4–6.

16.   Koert van Mensvoort and Hendrik-Jan Grievink, eds., *Next Nature: Nature Changes Along with Us* (Amsterdam: Next Nature Network, 2015), 275.

17.   Van Mensvoort and Grievink, *Next Nature*, 275.

18.   Linda J. Barth, *A History of Inventing in New Jersey: From Thomas Edison to the Ice Cream Cone* (Charleston, SC: History Press, 2013), 70.

19.   Elizabeth Fee and Theodore M. Brown, "John Harvey Kellogg, MD: Health Reformer and Antismoking Crusader," *American Journal of Public Health* 92, no. 6 (2002): 935.

20.   Smith, *Encyclopedia of Junk Food*, xxxiii, 22, 56, 79, 212.

21.   Smith, *Encyclopedia of Junk Food*, 53.

22.   Wiss, Avena, and Rada, "Sugar Addiction."

23.   Brand Finance, *Food & Drink 2019: The Annual Report on the Most Valuable and Strongest Food and Soft Drink Brand* (Brand Finance, 2019), 8.

24.   Oganisation for Economic Co-operation and Development (OECD), *OECD Environmental Outlook to 2050: The Consequences of Inaction* (Paris: OECD, 2012).

25.   Tim Smedley, "Is the World Running out of Fresh Water?," BBC, April 12, 2017, https://www.bbc.com/future/article/20170412-is-the-world-running-out-of-fresh-water; Alexandra S. Richey et al., "Quantifying Renewable Groundwater Stress with Grace," *Water Resources Research* 51, no. 7 (2015): 5224, 5231–5232.

26.   World Health Organization (WHO), "Drinking-Water," WHO, 2019, https://www.who.int/news-room/fact-sheets/detail/drinking-water.

27.   Xavier Leflaive et al., "Water," in *OECD Environmental Outlook to 2050: The Consequences of Inaction*, ed. Organisation for Economic Co-operation and Development (OECD) (Paris: OECD Publishing, 2012), 216.

28.   Helmut Smits, *The Real Thing*, 2010–2017, accessed September 2, 2021, http://helmutsmits.nl/work/real-thing.

29.   A. M. Al-Sabagha et al., "Greener Routes for Recycling of Polyethylene Terephthalate," *Egyptian Journal of Petroleum* 25, no. 1 (2016): 54.

30. Luis Marques, "Plasticene: The World as a Continuum of Polymers," in *Capitalism and Environmental Collapse* (Cham, Switzerland: Springer International, 2020), 106.

31. Luz Claudio, "Our Food: Packaging & Public Health," *Environmental Health Perspectives* 120, no. 6 (2012): 233–337.

32. Kai Olson-Sawyer and Robin Madel, "The Water Footprint of Your Plastic Bottle," Food Print, July 17, 2020, https://foodprint.org/blog/plastic-water-bottle/.

33. Shalini Tandon, Niranjan Kolekar, and Rakesh Kumar, "Water and Energy Footprint Assessment of Bottled Water Industries in India," *Natural Resources* 5 (2014): 71.

34. Peter Nowak, *Sex, Bombs, and Burgers: How War, Pornography, and Fast Food Have Shaped Modern Technology* (Guilford, CT: Lyons Press, 2010), 83.

35. Lisa R. Young and Marion Nestle, "Portion Sizes and Obesity: Responses of Fast-Food Companies," *Journal of Public Health Policy* 28, no. 2 (2007): 238–248.

36. Young and Nestle, "Portion Sizes and Obesity," 283.

37. Joel Fuhrman, "The Hidden Dangers of Fast and Processed Food," *American Journal of Lifestyle Medicine* 12, no. 5 (2018): 378.

38. Warren James Belasco, *Meals to Come: A History of the Future of Food* (Berkeley: University of California Press, 2006), 251.

39. Dale L. Huffman, J. A. Marchello, and T. P. Ringkob, "Development of Processed Meat Items for the Fast-Food Industry," *Reciprocal Meat Conference Proceedings* 39 (1986): 25–26.

40. Liming Wang and John Davis, *China's Grain Economy: The Challenge of Feeding More than a Billion* (London: Routledge, 2018), 140–141.

41. Yu Cui and Zhang Ting, "American Fast Food in Chinese Market: A Cross-Cultural Perspective—The Case of KFC and McDonald's," master's thesis, University of Halmstad, 2009, 22.

42. An Pan, Vasanti Malik, and Frank B. Hu, "Exporting Diabetes to Asia: The Impact of Western-Style Fast Food," *Circulation* 10, no. 126 (2012): 163–165.

43. Space10, *Tomorrow's Meatball: A Visual Exploration of Future Foods*, 2015, accessed August 13, 2021, https://space10.com/project/tomorrows-meatball/.

44. Karen Miner, "What You Don't Know about the IKEA Meatballs," Mashed, last updated March 22, 2021, https://www.mashed.com/99142/dont-know-ikea-meatballs/.

45. Space10, *Tomorrow's Meatball*.

46. Space10, *Tomorrow's Meatball*.

47. Michael Pollan, *The Omnivore's Dilemma: A Natural History of Four Meals* (New York: Penguin Books, 2006), 114.

48. Pollan, *The Omnivore's Dilemma*, 111.

49. Rose Etherington, "Slim Chips by Hafsteinn Juliusson," dezeen, April 19, 2010, https://www.dezeen.com/2010/04/19/slim-chips-by-hafsteinn-juliusson/.

50. Etherington, "Slim Chips."

51. Boo Chapple, *Consumables*, 2009, http://residualsoup.org/on-the-boil/consumables.html (site discontinued).

52. Chapple, *Consumables*.

53. Henry Hargreaves, *Deep Fried Gadgets*, 2013, accessed August 13, 2021, https://henryhargreaves.com/Deep-Fried-Gadgets.

54. Nicola Twilley, *Edible Cellphones*, October 14, 2009, https://www.ediblegeography.com/edible-cellphones (site discontinued); Hargreaves, *Deep Fried Gadgets*.

55. Starbucks, "Starbucks Weaves Its Magic with New Color and Flavor Changing Unicorn Frappuccino," Starbucks Stories & News, April 18, 2017, https://stories.starbucks.com/stories/2017/starbucks-unicorn-frappuccino.

Chapter 9

1. Michiel Korthals, *Before Dinner: Philosophy and Ethics of Food*, trans. Frans Kooymans (Dordrecht: Springer, 2004), 4–5.

2. Jessica R. Glass et al., "Was Frozen Mammoth or Giant Ground Sloth Served for Dinner at the Explorers Club?," *PLoS ONE* 11, no. 2 (2016): e0146825.

3. Glass et al., "Was Frozen Mammoth."

4. Lucia Martinelli, Markku Oksanen, and Helena Siipi, "De-Extinction: A Novel and Remarkable Case of Bio-Objectification," *Croatian Medical Journal* 55, no. 4 (2014): 423–427.

5. Martinelli, Oksanen, and Siipi, "De-Extinction," 423.

6. Martinelli, Oksanen, and Siipi, "De-Extinction," 423.

7. The Center for Genomic Gastronomy, *The De-Extinction Deli*, 2013, accessed September 8, 2021, https://genomicgastronomy.com/work/2013-2/deli.

8. Center for Genomic Gastronomy, *The De-Extinction Deli*.

9. Center for Genomic Gastronomy, *The De-Extinction Deli*.

10. S. T. Turvey and C. L. Risley, "Modelling the Extinction of Steller's Sea Cow," *Biology Letters* 2, no. 1 (2006): 94–97.

11. Yummy Dino Buddies, "Discover Kid-Friendly Favorites That Pack a Nutritional Punch," Maxi Canada, accessed September 7, 2021, https://www.yummydinobuddies.com/products; Van Mensvoort and Grievink, *Next Nature*, 313.

12. Anna V. Vitkalova et al., "Transboundary Cooperation Improves Endangered Species Monitoring and Conservation Actions: A Case Study of the Global Population of Amur Leopards," *Conservation Letters* 11, no. 5 (2018): 5, e12574.

13. Anthropocene Kitchen, *Anthropocene Feast: Eating the Last Dragon*, 2017, accessed March 5, 2022, https://anthropocenekitchen.wordpress.com/portfolio/eating-the-last-dragon.

14. Anthropocene Kitchen, *Anthropocene Feast: Eating the Last Dragon*.

15. Mindy Weisberger, "'Cosmic Mudball Meteorite' Smells Like Brussels Sprouts, Finds New Home at Museum," Live Science, October 10, 2019, https://www.livescience.com/mudball-meteorite-smells-like-brussels-sprouts.html.

16. Weisberger, "'Cosmic Mudball Meteorite' Smells Like Brussels Sprouts."

17. Tony Phillips, "The Mysterious Smell of Moondust," NASA Science, January 30, 2006, https://science.nasa.gov/science-news/science-at-nasa/2006/30jan_smellofmoondust.

18. Steven Spielberg, dir., *E.T. the Extra-Terrestrial* (Universal City, CA: Amblin Productions, 1982), film.

19. Erik Sandelin and Unsworn Industries, "Eating E.T.," in *Exploring the Animal Turn: Human-Animal Relations in Science, Society and Culture*, ed. Erika Andersson Cederholm et al. (Lund, Sweden: Pufendorfinstitutet, 2014), 47.

20. Sandelin and Unsworn Industries, "Eating E.T," 50.

21. Sandelin and Unsworn Industries, "Eating E.T," 52.

Chapter 10

1. Chris D. Thomas, "The Sixth Mass Genesis? New Species Are Coming into Existence Faster than Ever Thanks to Humans," The Conversation, October 6, 2019, https://theconversation.com/the-sixth-mass-genesis-new-species-are-coming-into-existence-faster-than-ever-thanks-to-humans-80527.

2. Thomas, "The Sixth Mass Genesis?"

3. Thomas, "The Sixth Mass Genesis?"

4. Slavoj Žižek, "Slavoj Žižek: 'Humanity Is OK, but 99% of People Are Boring Idiots,'" interview by Decca Aitkenhead, *The Guardian*, June 10, 2012.

5. Good Meat, "This Is GOOD Meat," Good Meat home page, accessed August 26, 2021, https://goodmeat.co.

6. Space10, "About," Space10, accessed July 28, 2021, https://space10.com/about.

7. Hartmann Schedel and Anton Koberger, *The Nuremberg Chronicle: A Facsimile of Hartmann Schedel's* Buch Der Chroniken, *Printed by Anton Koberger in 1493* (New York: Landmark Press, 1979).

8. Julian Cribb, *The Coming Famine: The Global Food Crisis and What We Can Do to Avoid It* (Berkeley: University of California Press, 2010), 187.

Selected Bibliography

Aday, Serpil, and Mehmet Seckin Aday. "Impact of Covid-19 on the Food Supply Chain." *Food Quality and Safety* 4, no. 4 (2020): 167–180. https://doi.org/10.1093/fqsafe/fyaa024.

Alexandratos, Nikos, and Jelle Bruinsma. "World Agriculture towards 2030/2050: The 2012 Revision." ESA Working Papers 288998, Food and Agriculture Organization of the United Nations, Agricultural Development Economics Division, 2012. http://dx.doi.org/10.22004/ag.econ.288998.

Al-Sabagha, A. M., F. Z. Yehia, Gh Eshaq, A. M. Rabie, and A. E. El-Metwally. "Greener Routes for Recycling of Polyethylene Terephthalate." *Egyptian Journal of Petroleum* 25, no. 1 (2016): 53–64. https://doi.org/10.1016/j.ejpe.2015.03.001.

Andrade, Cristiane C. P., Fernanda Mandelli, Sergio Echeverrigaray, and Ana P. L. Delamare. "Microbial Dynamics during Cheese Production and Ripening: Physicochemical and Biological Factors." *Food Global Science Books* 2, no. 2 (2008): 91–101.

Andrews, Lori. "Tissue Culture: The Line between Art and Science Blurs When Two Artists Hang Cells in Galleries." *Journal of Life Sciences* (September 2007): 68–73.

Arias-Maldonado, Manuel. "The Anthropocenic Turn: Theorizing Sustainability in a Postnatural Age." *Sustainability* 8, no. 1 (2015): 10. https://doi.org/10.3390/su8010010.

Ashton, Kevin. *How to Fly a Horse: The Secret History of Creation, Invention and Discovery*. New York: Anchor Books, 2015.

Baratz, Amit. "The Source of the Gods' Immortality in Archaic Greek Literature." *Scripta Classica Israelica* 34 (2015): 151–164.

Barkia, Ines, Nazamid Saar, and Schonna R. Manning. "Microalgae for High-Value Products towards Human Health and Nutrition." *Marine Drugs* 17, no. 5 (2019): 304. https://doi.org/10.3390/md17050304.

Barnosky, Anthony D., Nicholas Matzke, Susumu Tomiya, Guinevere O. U. Wogan, Brian Swartz, Tiago B. Quental, Charles Marshall, et al. "Has the Earth's Sixth Mass Extinction Already Arrived?" *Nature* 471, no. 7336 (2011): 51–57. https://doi.org/10.1038/nature09678.

Barth, Linda J. *A History of Inventing in New Jersey: From Thomas Edison to the Ice Cream Cone*. Charleston, SC: History Press, 2013.

Baum, Seth D., David C. Denkenberger, Joshua M. Pearce, Alan Robock, and Richelle Winkler. "Resilience to Global Food Supply Catastrophes." *Environment Systems and Decisions* 35, no. 2 (2015): 301–313. https://doi.org/10.1007/s10669-015-9549-2.

Bawa, A. S., and K. R. Anilakumar. "Genetically Modified Foods: Safety, Risks and Public Concerns—A Review." *Journal of Food Science and Technology* 50, no. 6 (2013): 1035–1046. https://doi.org/10.1007/s13197-012-0899-1.

Belasco, Warren James. "Future Notes: The Meal-in-a-Pill." In *Food in the USA: A Reader*, edited by Carole M. Counihan, 59–74. New York: Routledge, 2002.

Belasco, Warren James. *Meals to Come: A History of the Future of Food.* Berkeley: University of California Press, 2006.

Benke, Kurt, and Bruce Tomkins. "Future Food-Production Systems: Vertical Farming and Controlled-Environment Agriculture." *Sustainability: Science, Practice and Policy* 13, no. 1 (2017): 13–26. https://doi.org/10.1080/15487733.2017.1394054.

Bennett, Carys E., Richard Thomas, Mark Williams, Jan Zalasiewicz, Matt Edgeworth, Holly Miller, Ben Coles, Alison Foster, Emily J. Burton, and Upenyu Marume. "The Broiler Chicken as a Signal of a Human Reconfigured Biosphere." *Royal Society Open Science* 5, no. 12 (2018): 180325. https://doi.org/10.1098/rsos.180325.

Blazenhoff, Rusty. "An Edible Desert Survival Manual by Land Rover." Laughing Squid. May 10, 2012. https://laughingsquid.com/an-edible-desert-survival-manual-by-land-rover/.

Boje, David M. *Storytelling in the Global Age: There Is No Planet B.* Hackensack, NJ: World Scientific, 2019.

Boyd-Orr, John, and David Lubbock. *The White Man's Dilemma: Food and the Future.* 2nd ed. London: George Allen and Unwin, 1965. First published in 1953.

Brand Finance. *Food & Drink 2019: The Annual Report on the Most Valuable and Strongest Food and Soft Drink Brand.* Brand Finance, 2019.

Bratman, Steven. "Orthorexia vs. Theories of Healthy Eating." *Eating and Weight Disorders* 22 (2017): 381–385. https://doi.org/10.1007/s40519-017-0417-6.

Breslin, Paul A. S. "An Evolutionary Perspective on Food and Human Taste." *Current Biology* 23, no. 9 (2013): 409–418. https://doi.org/10.1016/j.cub.2013.04.010.

Brillat-Savarin, Jean Anthelme. *The Physiology of Taste: Or Meditations on Transcendental Gastronomy.* Translated by M. F. K. Fisher. New York: Everyman's Library, 2009. First published 1949 by the Heritage Press (New York).

Brindley, David A., Natasha L. Davie, Emily J. Culme-Seymour, Chris Mason, David W. Smith, and Jon A. Rowley. "Peak Serum: Implications of Serum Supply for Cell Therapy Manufacturing." *Future Medicine* 7, no. 1 (2012): 7–13. https://doi.org/10.2217/RME.11.112.

Bruinsma, Jelle. "The Resources Outlook: By How Much Do Land, Water and Crop Yields Need to Increase by 2050?" In *Looking Ahead in World Food and Agriculture: Perspectives to 2050*, edited by Piero Conforti. Rome: FAO, 2011.

Cadinu, Lorenzo A., Paolo Barra, Francesco Torre, Francesco Delogu, and Fabio A. Madau. "Insect Rearing: Potential, Challenges, and Circularity." *Sustainability* 12, no. 11 (2020): 4567. https://doi.org/10.3390/su12114567.

Čadová, Michaela, Renata Havránková, Jiří Havránek, and Friedo Zölzer. "Radioactivity in Mushrooms from Selected Locations in the Bohemian Forest, Czech Republic." *Radiation and Environmental Biophysics* 56, no. 2 (2017): 167–175. https://doi.org/10.1007/s00411-017-0684-7.

Catts, Oron, and Ionat Zurr. "Disembodied Livestock: The Promise of a Semi-Living Utopia." *Parallax* 19, no. 1 (2013): 101–113. https://doi.org/10.1080/13534645.2013.752062.

Ceballos, Gerardo, Paul R. Ehrlich, and Rodolfo Dirzo. "Biological Annihilation via the Ongoing Sixth Mass Extinction Signaled by Vertebrate Population Losses and Declines." *PNAS* 114, no. 30 (2017): 6089–6096. https://doi.org/10.1073/pnas.1704949114.

Çelik, Özge, and Çimen Atak. "Applications of Ionizing Radiation in Mutation Breeding." In *New Insights on Gamma Rays*, edited by Ahmed M. Maghraby, 111–132. London: InTech, 2017.

Cerritos, René. "Insects as Food: An Ecological, Social and Economical Approach." *CAB Reviews: Perspectives in Agriculture Veterinary Science Nutrition and Natural Resources* 4, no. 27 (2009). https://doi.org/10.1079/PAVSNNR20094027.

Chela-Flores, Julian. "Terrestrial Microbes as Candidates for Survival on Mars and Europa." In *Journey to Diverse Microbial Worlds: Adaptation to Exotic Environments*, edited by Joseph Seckbach, 387–398. Dordrecht: Springer, 2000.

Chitchumroonchokchai, Chureeporn, Gianfranco Diretto, Bruno Parisi, Giovanni Giuliano, and Mark L. Failla. "Potential of Golden Potatoes to Improve Vitamin A and Vitamin E Status in Developing Countries." *PLoS One* 12, no. 11 (2017): e0187102. https://doi.org/10.1371/journal.pone.0187102.

Churchill, Winston. "Fifty Years Hence." *Strand Magazine*, December 1931, 200.

Claudio, Luz. "Our Food: Packaging & Public Health." *Environmental Health Perspectives* 120, no. 6 (2012): 233–237. https://doi.org/10.1289/ehp.120-a232.

Cohen, Mathilde. "The Lactating Man." In *Making Milk: The Past, Present, and Future of Our Primary Food*, edited by Mathilde Cohen and Yoriko Otomo, 141–160. London: Bloomsbury Academic, 2017.

Cole, James. "Assessing the Calorific Significance of Episodes of Human Cannibalism in the Palaeolithic." *Scientific Reports* 7, no. 1 (2017). https://doi.org/10.1038/srep44707.

Cooley, John R., David C. Marshall, Chris Simon, Michael L. Neckermann, and Gerry Bunker. "At the Limits: Habitat Suitability Modelling of Northern 17-Year Periodical Cicada Extinctions (*Hemiptera: Magicicada Spp.*)." *Global Ecology and Biogeography* 22, no. 4 (2013): 410–421. https://doi.org/10.1111/geb.12002.

Corcoran, Patricia L., Charles J. Moore, and Kelly Jazvac. "An Anthropogenic Marker Horizon in the Future Rock Record." *GSA Today* 24, no. 6 (2014): 4–8. https://doi.org/10.1130/gsat-g198a.1.

Cotton, William R., and Roger A. Pielke. "Nuclear Winter." In *Human Impacts on Weather and Climate*, 2nd ed., 203–219. Cambridge: Cambridge University Press, 2008.

Cribb, Julian. *The Coming Famine: The Global Food Crisis and What We Can Do to Avoid It*. Berkeley: University of California Press, 2010.

Crutzen, Paul J., and Eugene F. Stoermer. "The 'Anthropocene.'" *Global Change Newsletter* 41 (2000): 17–18.

Cui, Yu, and Zhang Ting. "American Fast Food in Chinese Market: A Cross-Cultural Perspective—The Case of KFC and McDonald's." Master's thesis, University of Halmstad, 2009.

Datta, Karabi, Gayetri Sahoo, Sellappan Krishnan, Moumita Ganguly, and Swapan K. Datta. "Genetic Stability Developed for β-Carotene Synthesis in BR29 Rice Line Using Dihaploid Homozygosity." *PLoS ONE* 9, no. 6 (2014). https://doi.org/10.1371/journal.pone.0100212.

De Chirico, Giorgio. "Meditations of a Painter: What the Painting of the Future Might Be, 1912." Translated by Louis Bourgeois and Robert Goldwater. In *Theories of Modern Art: A Source Book of Artists and Critics*. Edited by Herschel B. Chipp. Berkeley: University of California Press, 1996.

De Gramont, Sanche. "Popular Cheese in Italy Exposed as Garbage." *Washington Post*, September 20, 1962.

De Waal, Alex. *Mass Starvation the History and Future of Famine*. Cambridge: Polity Press, 2018.

Debackere, Boris, Andreas Eggertsen Tender, Carsten Friberg, Elizabeth Jochum, Michelle Kasprzak, Jakob Borrits Sabra, and Stahl Stenslie. *Innovation in Extreme Scenarios*. Rotterdam: V2_ Institute for the Unstable Media, 2014. https://v2.nl/archive/articles/innovation-in-extreme-scenarios-ebook.

Dell'Apa, Andrea, M. Chad Smith, and Mahealani Y. Kaneshiro-Pineiro. "The Influence of Culture on the International Management of Shark Finning." *Environmental Management* 54, no. 2 (2014): 151–161. https://doi.org/10.1007/s00267-014-0291-1.

Denfeld, Zack, and Cathrine Kramer. "Feeding Dangerous Ideas." Interview by Zane Cerpina. *EE Experimental Emerging Art* no. 3 (2018): 46–49.

Denkenberger, David Charles, and Joshua Pearce. *Feeding Everyone No Matter What: Managing Food Security after Global Catastrophe*. London: Academic Press, 2015.

Diamond, Jared. *Collapse: How Societies Choose to Fail or Survive*. New York: Viking, 2005.

DK. *The Story of Food: An Illustrated History of Everything We Eat*. Foreword by Giles Coren. New York: DK Publishing, 2018.

Dodd, Anna Bowman. *The Republic of the Future: Or, Socialism a Reality*. New York: Cassell & Company, 1887.

Drucker, Dorothée G., Yuichi I. Naito, Stéphane Péan, Sandrine Prat, Laurent Crépin, Yoshito Chikaraishi, Naohiko Ohkouchi, et al. "Isotopic Analyses Suggest Mammoth and Plant in the Diet of the Oldest Anatomically Modern Humans from Far Southeast Europe." *Scientific Reports* 7, no. 1 (2017). https://doi.org/10.1038/s41598-017-07065-3.

Dubbeling, Marielle, and Henk de Zeeuw. "Urban Agriculture and Climate Change Adaptation: Ensuring Food Security through Adaptation." In *Resilient Cities: Cities and Adaptation to Climate Change—Proceedings of the Global Forum 2010*, edited by Konrad Otto-Zimmermann, 441–450. Dordrecht: Springer, 2011.

Dunkel, F. V., and C. Payne. "Introduction to Edible Insects." In *Insects as Sustainable Food Ingredients: Production, Processing and Food Applications*, edited by Aaron T. Dossey, Juan A. Morales-Ramos, and M. Guadalupe Rojas, 1–27. Amsterdam: Academic Press, 2016.

Dunne, Anthony, and Fiona Raby. *Speculative Everything: Design, Fiction, and Social Dreaming*. Cambridge, MA: MIT Press, 2013.

Durso, L., and R. Hutkins. "Starter Cultures." In *Encyclopedia of Food Sciences and Nutrition*, 2nd ed., edited by Benjamin Caballero, 5583–5593. San Diego: Academic Press, 2003.

Edwards, Jesse J., Sandra L. Tollaksen, and Norman G. Anderson. "Proteins of Human Semen. I. Two-Dimensional Mapping of Human Seminal Fluid." *Clinical Chemistry* 27, no. 8 (1981): 1335–1340.

Ellis, Chris. "The Noah Virus: Who Is Infected with High Resiliency for Disaster?" Unpublished manuscript, last modified March 15, 2021.

Elorinne, Anna-Liisa, Mari Niva, Outi Vartiainen, and Pertti Väisänen. "Insect Consumption Attitudes among Vegans, Non-Vegan Vegetarians, and Omnivores." *Nutrients* 11, no. 2 (2019): 292. https://doi.org/10.3390/nu11020292.

Erens, Jesse, Sam van Es, Fay Haverkort, Eleni Kapsomenou, and Andy Luijben. "A Bug's Life: Large-Scale Insect Rearing in Relation to Animal Welfare." VENIK, Wageningen, 2012.

European Parliament, and Council of the European Union. "Regulation (EU) No 1308/2013: Establishing a Common Organisation of the Markets in Agricultural Products and Repealing Council Regulations (EEC) No 922/72, (EEC) No 234/79, (EC) No 1037/2001 and (EC) No 1234/2007." *Official Journal of the European Union* (December 2013): 671–854.

Evans, Josh, Roberto Flore, Michael Bom Frøst, and Nordic Food Lab. *On Eating Insects: Essays, Stories and Recipes*. London: Phaidon Press, 2017.

Faulk, Richard. *The Next Big Thing: A History of the Boom-or-Bust Moments That Shaped the Modern World*. San Francisco: Zest Books, 2015.

Fee, Elizabeth, and Theodore M. Brown. "John Harvey Kellogg, MD: Health Reformer and Antismoking Crusader." *American Journal of Public Health* 92, no. 6 (2002): 935. https://doi.org/10.2105/AJPH.92.6.935.

Felbab-Brown, Vanda. *The Extinction Market: Wildlife Trafficking and How to Counter It*. Oxford: Oxford University Press, 2017.

Flidel-Rimon, O., and E. S. Shinwell. "Breast Feeding Twins and High Multiples." *Archives of Disease in Childhood: Fetal and Neonatal* 91, no. 5 (2006): 377–380. https://doi.org/10.1136/adc.2005.082305.

Food and Agriculture Organization of the United Nations (FAO). *Climate Change and Food Security: Risks and Responses*. Rome: FAO, 2015.

Food and Agriculture Organization of the United Nations (FAO). *Desert Locust Crisis: Appeal for Rapid Response and Anticipatory Action in the Greater Horn of Africa*. Rome: FAO, 2020.

Fraser, Evan D. G. "Social Vulnerability and Ecological Fragility: Building Bridges between Social and Natural Sciences Using the Irish Potato Famine as a Case Study." *Conservation Ecology* 7, no. 2 (2003). https://doi.org/10.5751/ES-00534-070209.

Fuhrman, Joel. "The Hidden Dangers of Fast and Processed Food." *American Journal of Lifestyle Medicine* 12, no. 5 (2018): 375–381. https://doi.org/10.1177/1559827618766483.

Gale, Fred, and Jean C. Buzby. "Imports from China and Food Safety Issues." *United States Department of Agriculture, Economic Research Service, Economic Information Bulletin*, July 2009.

Garnett, Tara. "Plating Up Solutions." *Science* 353, no. 6305 (2016): 1202–1204. https://doi.org/10.1126/science.aah4765.

Gasperi, Johnny, Stephanie L. Wright, Rachid Dris, France Collard, Corinne Mandin, Mohamed Guerrouache, Valérie Langlois, Frank J. Kelly, and Bruno Tassin. "Microplastics in Air: Are We Breathing It In?" *Current Opinion in Environmental Science & Health* 1 (2018): 1–5. https://doi.org/10.1016/j.coesh.2017.10.002.

Gasteratos, Kristopher. "Nature & the Neomnivore." *The Cellular Agriculture Environmental Impact Compendium*, August 2017.

Geiger, Ernest. "Problems Connected with the Possible Use of Plankton for Human Nutrition." *American Journal of Clinical Nutrition* 6, no. 4 (1958): 394–400. https://doi.org/10.1093/ajcn/6.4.394.

Gerbault, Pascale, Anke Liebert, Yuval Itan, Adam Powell, Mathias Currat, Joachim Burger, Dallas M. Swallow, and Mark G. Thomas. "Evolution of Lactase Persistence: An Example of Human Niche Construction." *Philosophical Transactions of the Royal Society B: Biological Sciences* 366, no. 1566 (2011): 863–877. https://doi.org/10.1098/rstb.2010.0268.

Gere, Attila, Ryan Zemel, Dalma Radványi, and Howard Moskowitz. "Insect Based Foods: A Nutritional Point of View." *Nutrition & Food Science International Journal* 4, no. 3 (2017): 555638. https://doi.org/10.19080/nfsij.2017.04.555638.

Gholamipour-Shirazi, Azarmidokht, Michael-Alex Kamlow, Ian T. Norton, and Tom Mills. "How to Formulate for Structure and Texture via Medium of Additive Manufacturing: A Review." *Foods* 9, no. 4 (2020): 497. https://doi.org/10.3390/foods9040497.

Glass, Jessica R., Matt Davis, Timothy J. Walsh, Eric J. Sargis, and Adalgisa Caccone. "Was Frozen Mammoth or Giant Ground Sloth Served for Dinner at the Explorers Club?" *PLoS One* 11, no. 2 (2016): e0146825. https://doi.org/10.1371/journal.pone.0146825.

Godfray, H. C. J., Paul Aveyard, Tara Garnett, Jim W. Hall, Timothy J. Key, Jamie Lorimer, Ray T. Pierrehumbert, Peter Scarborough, Marco Springmann, and Susan A. Jebb. "Meat Consumption, Health, and the Environment." *Science* 361, no. 6399 (2018): 1–8. https://doi.org/10.1126/science.aam5324.

Godfray, H. C. J., and Oxford Martin School, Oxford University. "Meat: The Future Series—Alternative Proteins." World Economic Forum, 2019. https://eprints.whiterose.ac.uk/170474.

Good Meat. "This Is GOOD Meat." Good Meat. Accessed August 26, 2021. https://goodmeat.co/.

Goulson, Dave. "The Insect Apocalypse, and Why It Matters." *Current Biology* 29, no. 19 (2019). https://doi.org/10.1016/j.cub.2019.06.069.

Gräslund, Bo, and Neil Price. "Twilight of the Gods? The 'Dust Veil Event' of AD 536 in Critical Perspective." *Antiquity* 86, no. 332 (2012): 428–443. https://doi.org/10.1017/s0003598x00062852.

Greville, Warwick L. "Proposed Safety Guidelines for the Maximum Volume Fat Removal by Tumescent Liposuction." In *Liposuction*, edited by Melvin A. Shiffman and Alberto Di Giuseppe, 30–35. Berlin: Springer, 2006.

Groeniger, Joost Oude, Frank J. van Lenthe, Mariëlle A. Beenackers, and Carlijn B. M. Kamphuis. "Does Social Distinction Contribute to Socioeconomic Inequalities in Diet: The Case of 'Superfoods' Consumption." *International Journal of Behavioral Nutrition and Physical Activity* 14, no. 1 (2017): 40. https://doi.org/10.1186/s12966-017-0495-x.

Gustavsson, Jenny, Alexandre Meybeck, Robert van Otterdijk, Ulf Sonesson, and Christel Cederberg. *Global Food Losses and Food Waste: Extent, Causes and Prevention.* Rome: Food and Agriculture Organization of the United Nations, 2011.

Gutierrez, Andrew Paul, and Luigi Ponti. "Analysis of Invasive Insects: Links to Climate Change." In *Invasive Species and Global Climate Change*, edited by Lewis H. Ziska and Jeffrey S. Dukes, 45–61. Wallingford, UK: CABI, 2014.

Haas, Rainer, Alina Schnepps, Anni Pichler, and Oliver Meixner. "Cow Milk versus Plant-Based Milk Substitutes: A Comparison of Product Image and Motivational Structure of Consumption." *Sustainability* 11, no. 18 (2019): 5046. https://doi.org/10.3390/su11185046.

Harari, Yuval Noah. *Homo Deus: A Brief History of Tomorrow*. New York: Harper Perennial, 2016.

Hartmann, Christina, Jing Shi, Alice Giusto, and Michael Siegrist. "The Psychology of Eating Insects: A Cross-Cultural Comparison between Germany and China." *Food Quality and Preference* 44 (2015): 148–156. https://doi.org/10.1016/j.foodqual.2015.04.013.

Henchion, Maeve, Maria Hayes, Anne Mullen, Mark Fenelon, and Brijesh Tiwari. "Future Protein Supply and Demand: Strategies and Factors Influencing a Sustainable Equilibrium." *Foods* 6, no. 7 (2017): 53. https://doi.org/10.3390/foods6070053.

Herrero, Mario, Stefan Wirsenius, Benjamin Henderson, Cyrille Rigolot, Philip Thornton, Petr Havlík, Imke De Boer, and Pierre J. Gerber. "Livestock and the Environment: What Have We Learned in the Past Decade?" *Annual Review of Environment and Resources* 40, no. 1 (2015): 177–202. https://doi.org/10.1146/annurev-environ-031113-093503.

Hinde, Katie, and Lauren A. Milligan. "Primate Milk: Proximate Mechanisms and Ultimate Perspectives." *Evolutionary Anthropology: Issues, News, and Reviews* 20, no. 1 (2011): 9–23. https://doi.org/10.1002/evan.20289.

Hoeling, Barbara, Douglas Reed, and P. B. Siegel. "Going Bananas in the Radiation Laboratory." *American Journal of Physics* 67, no. 5 (1999): 440–442. https://doi.org/10.1119/1.19281.

Hosen, S. M. Zahid, Swati Paul, and Dibyajyoti Saha. "Artificial and Fake Eggs: Dance of Death." *Advances in Pharmacology and Pharmacy* 1, no. 1 (2013): 13–17.

Houellebecq, Michael. *The Possibility of an Island*. Translated by Gavin Bowd. 2005. New York: Vintage Books, 2007.

Hubert, Ph., F. Perrot, J. Gaye, B. Médina, and M. S. Pravikoff. "Radioactivity Measurements Applied to the Dating and Authentication of Old Wines." *Comptes Rendus Physique* 10, no. 7 (2009): 622–629. https://doi.org/10.1016/j.crhy.2009.08.007.

Huffman, Dale L., J. A. Marchello, and T. P. Ringkob. "Development of Processed Meat Items for the Fast-Food Industry." *Reciprocal Meat Conference Proceedings* 39 (1986): 25–26.

Hume, Stephen, Alexandra Morton, Betty Keller, Rosella M. Leslie, Otto Langer, and Don Staniford. *A Stain upon the Sea: West Coast Salmon Farming*. Madeira Park, BC: Harbour Publishing, 2004.

Hutchings, Jeffrey A., and John D. Reynolds. "Marine Fish Population Collapses: Consequences for Recovery and Extinction Risk." *BioScience* 54, no. 4 (2004): 297–309. https://doi.org/dqfb9z.

Imathiu, Samuel. "Benefits and Food Safety Concerns Associated with Consumption of Edible Insects." *NFS Journal* 18 (2020): 1–11. https://doi.org/10.1016/j.nfs.2019.11.002.

International Society of Aesthetic Plastic Surgery (ISAPS). *ISAPS International Survey on Aesthetic/Cosmetic Procedures Performed in 2019*, 2020.

Ito, Hiromu, Satoshi Kakishima, Takashi Uehara, Satoru Morita, Takuya Koyama, Teiji Sota, John R. Cooley, and Jin Yoshimura. "Evolution of Periodicity in Periodical Cicadas." *Scientific Reports* 5, no. 1 (2015): 14094. https://doi.org/10.1038/srep14094.

Jaouen, Klervia, Michael P. Richards, Adeline Le Cabec, Frido Welker, William Rendu, Jean-Jacques Hublin, Marie Soressi, and Sahra Talamo. "Exceptionally High δ15N Values in Collagen Single Amino Acids Confirm Neandertals as High-Trophic Level Carnivores." *Proceedings of the National Academy of Sciences* 116, no. 11 (2019): 4928–4933. https://doi.org/10.1073/pnas.1814087116.

Jung, F., A. Krüger-Genge, P. Waldeck, and J. H. Küpper. "*Spirulina platensis*, a Super Food?" *Journal of Cellular Biotechnology* 5, no. 1 (2019): 43–54. https://doi.org/10.3233/jcb-189012.

Kanter, Mitch, and Ashley Desrosiers. "Personalized Wellness Past and Future: Will the Science and Technology Coevolve?" *Nutrition Today* 54, no. 4 (2019): 174–181. https://doi.org/10.1097/nt.0000000000000354.

Karami, Ali, Abolfazl Golieskardi, Cheng Keong Choo, Vincent Larat, Tamara S. Galloway, and Babak Salamatinia. "The Presence of Microplastics in Commercial Salts from Different Countries." *Scientific Reports* 7, no. 1 (2017): 46173. https://doi.org/10.1038/srep46173.

Karatzas, Costas N., and Jeffrey D. Turner. "Toward Altering Milk Composition by Genetic Manipulation: Current Status and Challenges." *Journal of Dairy Science* 80, no. 9 (1997): 2225–2232. https://doi.org/10.3168/jds.s0022-0302(97)76171-x.

Kauppi, Saara Maria. "Insect Economy and Marketing: How Much and in What Way Could Insects be Shown in Packaging?" Master's thesis, Aalto University, 2016. http://dx.doi.org/10.13140/RG.2.2.15776.20486.

Kauppi, Saara-Maria. "Packaging Design Strategies for Introducing Whole Mealworms as Human Food." In *DS-101: Proceedings of NordDesign 2020, Lyngby, Denmark, 12th–14th August 2020—Balancing Innovation and Operation*, edited by N. H. Mortensen, C. T. Hansen, and M. Deininger. Design Society, 2020. https://doi.org/10.35199/norddesign2020.42.

Kauppi, Saara Maria, Ida Nilstad Pettersen, and Casper Boks. "Consumer Acceptance of Edible Insects and Design Interventions as Adoption Strategy." *International Journal of Food Design* 4, no. 1 (2019): 39–62. https://doi.org/10.1386/ijfd.4.1.39_1.

Kettenburg, Annika J., Jan Hanspach, David J. Abson, and Joern Fischer. "From Disagreements to Dialogue: Unpacking the Golden Rice Debate." *Sustainability Science* 13, no. 5 (2018): 1469–1482. https://doi.org/10.1007/s11625-018-0577-y.

Klein, Alexandra-Maria, Bernard E. Vaissière, James H. Cane, Ingolf Steffan-Dewenter, Saul A. Cunningham, Claire Kremen, and Teja Tscharntke. "Importance of Pollinators in Changing Landscapes for World Crops." *Proceedings of the Royal Society B: Biological Sciences* 274, no. 1608 (2006): 303–313. https://doi.org/10.1098/rspb.2006.3721.

Koelmans, Albert A., Nur Hazimah Mohamed Nor, Enya Hermsen, Merel Kooi, Svenja M. Mintenig, and Jennifer De France. "Microplastics in Freshwaters and Drinking Water: Critical Review and Assessment of Data Quality." *Water Research* 155 (2019): 410–422. https://doi.org/10.1016/j.watres.2019.02.054.

Korthals, Michiel. *Before Dinner: Philosophy and Ethics of Food*. Translated by Frans Kooymans. Dordrecht: Springer, 2004.

Kosuth, Mary, Sherri A. Mason, and Elizabeth V. Wattenberg. "Anthropogenic Contamination of Tap Water, Beer, and Sea Salt." *PLoS One* 13, no. 4 (2018): e0194970. https://doi.org/10.1371/journal.pone.0194970.

Kouřimská, Lenka, and Anna Adámková. "Nutritional and Sensory Quality of Edible Insects." *NFS Journal* 4 (2016): 22–26. https://doi.org/10.1016/j.nfs.2016.07.001.

Kozjak, Petra, and Vladimir Meglič. "Mutagenesis in Plant Breeding for Disease and Pest Resistance." In *Mutagenesis*, edited by Rajnikant Mishra, 195–220. Rijeka, Croatia: InTech, 2012.

Kulski, Jerzy K., and Peter E. Hartmann. "Composition of Breast Fluid of a Man with Galactorrhea and Hyperprolactinaemia." *Journal of Clinical Endocrinology & Metabolism* 52, no. 3 (1981): 581–582. https://doi.org/10.1210/jcem-52-3-581.

Larsson, Josefin. "Foods of the Future: Gastropolitics and Climate Change in the Anthropocene." Master's thesis, Department of Arts and Cultural Sciences, Lund University, 2018.

Lassa, Jonatan A., Paul Teng, Mely Caballero-Anthony, and Maxim Shrestha. "Revisiting Emergency Food Reserve Policy and Practice under Disaster and Extreme Climate Events." *International Journal of Disaster Risk Science* 10 (2019): 1–13. https://doi.org/10.1007/s13753-018-0200-y.

Lauriola, Rosanna. "The Greeks and the Utopia: An Overview through Ancient Greek Literature." *Revista Espaço Acadêmico* 9, no. 97 (2009): 109–124.

Lees, Alexander C., and Stuart L. Pimm. "Species, Extinct before We Know Them?" *Current Biology* 25, no. 5 (2015): 177–180. https://doi.org/10.1016/j.cub.2014.12.017.

Leflaive, Xavier, Kayoung Kim, Henk Hilderink, Arno Bouwman, Hans Visser, Lex Bouwman, Tom Kram, Marloes Bakke, Roberto Martin-Hurtado, and Maria Witmer. "Water." In *OECD Environmental Outlook to 2050: The Consequences of Inaction*, edited by Organisation for Economic Co-operation and Development (OECD), 207–274. Paris: OECD, 2012.

Liebezeit, Gerd, and Elisabeth Liebezeit. "Non-Pollen Particulates in Honey and Sugar." *Food Additives & Contaminants: Part A* 30, no. 12 (2014): 2136–2140. https://doi.org/10.1080/19440049.2013.843025.

Liebezeit, Gerd, and Elisabeth Liebezeit. "Synthetic Particles as Contaminants in German Beers." *Food Additives & Contaminants: Part A* 31, no. 9 (2014): 1574–1578. https://doi.org/10.1080/19440049.2014.945099.

Liebhold, Andrew M., Michael J. Bohne, and Rebecca L. Lilja. "Active Periodical Cicada Broods of the United States." USDA Forest Service Northern Research Station, Northeastern Area State & Private Forestry, 2013.

Linder, Tomas. "Making the Case for Edible Microorganisms as an Integral Part of a More Sustainable and Resilient Food Production System." *Food Security* 11, no. 2 (2019): 265–278. https://doi.org/10.1007/s12571-019-00912-3.

Livin Farms. *The Hive Magazine*, 2017.

Lobanovska, Mariya, and Giulia Pilla. "Penicillin's Discovery and Antibiotic Resistance: Lessons for the Future?" *Yale Journal of Biology and Medicine* 90, no. 1 (2017): 135–145.

Lu, Jiangyong, and Zhigang Tao. "Sanlu's Melamine-Tainted Milk Crisis in China." Asia Case Research Centre, University of Hong Kong, June 2009, 1–24.

Lundqvist, U. "Eighty Years of Scandinavian Barley Mutation Genetics and Breeding." In *Induced Plant Mutations in the Genomics Era*, edited by Q. Y. Shu, 39–43. Rome: FAO, 2009.

Madigan, D. J., Z. Baumann, and N. S. Fisher. "Pacific Bluefin Tuna Transport Fukushima-Derived Radionuclides from Japan to California." *Proceedings of the National Academy of Sciences* 109, no. 24 (2012): 9483–9486. https://doi.org/10.1073/pnas.1204859109.

Malpass, Matt. *Critical Design in Context: History, Theory, and Practice*. London: Bloomsbury Publishing, 2019.

Malthus, Thomas R. *An Essay on the Principle of Population, as It Affects the Future Improvement of Society; with Remarks on the Speculations of W. Godwin, M. Condorcet and Other Writers.* Bellingham, WA: Electronic Scholarly Publishing Project, 1998. First published in London in 1798 by J. Johnson.

Marinetti, Filippo Tommaso. *The Futurist Cookbook.* Edited by Lesley Chamberlain. Translated by Suzanne Brill. London: Penguin Books, 2014. First published in Italian in 1932 as *La Cucina Futurista.*

Marques, Luis. "Plasticene: The World as a Continuum of Polymers." In *Capitalism and Environmental Collapse*, 106–108. Cham, Switzerland: Springer International, 2020.

Martinelli, Lucia, Markku Oksanen, and Helena Siipi. "De-Extinction: A Novel and Remarkable Case of Bio-Objectification." *Croatian Medical Journal* 55, no. 4 (2014): 423–427. https://doi.org/10.3325/cmj.2014.55.423.

Martins, Maristela, Ariane Mendonça Pacheco, Ana Cyra Santos Lucas, Avacir Casanova Andrello, Carlos Roberto Appoloni, and Jose Junior Mendonça Xavier. "Brazil Nuts: Determination of Natural Elements and Aflatoxin." *Acta Amazonica* 42, no. 1 (2012): 157–164. https://doi.org/10.1590/s0044-59672012000100018.

McClements, David Julian, Emily Newman, and Isobelle Farrell McClements. "Plant-Based Milks: A Review of the Science Underpinning Their Design, Fabrication, and Performance." *Comprehensive Reviews in Food Science and Food Safety* 18, no. 6 (2019): 2047–2067. https://doi.org/10.1111/1541-4337.12505.

McDade, Harry, and C. Matilda Collins. "How We Might Overcome 'Western' Resistance to Eating Insects." In *Edible Insects*, edited by Heimo Mikkola, 5–16. London: IntechOpen, 2020.

Mead, Simon, Michael P. H. Stumpf, Jerome Whitfield, Jonathan A. Beck, Mark Poulter, Tracy Campbell, James Uphill, et al. "Balancing Selection at the Prion Protein Gene Consistent with Prehistoric Kurulike Epidemics." *Science* 300, no. 5619 (2003): 640–643. https://doi.org/10.1126/science.1083320.

Megginson, Leon C. "Lessons from Europe for American Business." *Southwestern Social Science Quarterly* 44, no. 1 (1963): 3–13.

Miglietta, Pier, Federica De Leo, Marcello Ruberti, and Stefania Massari. "Mealworms for Food: A Water Footprint Perspective." *Water* 7, no. 11 (2015): 6190–6203. https://doi.org/10.3390/w7116190.

Miller, Toby. *Greenwashing Culture.* Abington, UK: Routledge, 2018.

Morton, Timothy, ed. *Cultures of Taste / Theories of Appetite: Eating Romanticism.* New York: Palgrave Macmillan, 2004.

Morton, Timothy. *The Ecological Thought.* Cambridge, MA: Harvard University Press, 2012.

Morton, Timothy. *Ecology without Nature: Rethinking Environmental Aesthetics.* Cambridge, MA: Harvard University Press, 2007.

Morton, Timothy. *Hyperobjects: Philosophy and Ecology after the End of the World.* Minneapolis: University of Minnesota Press, 2017.

Mouritsen, Ole G., and Klavs Styrbæk. *Mouthfeel: How Texture Makes Taste.* Translated by Mariela Johansen. New York: Columbia University Press, 2017.

Mühlschlegel, Peter, Armin Hauk, Ulrich Walter, and Robert Sieber. "Lack of Evidence for Microplastic Contamination in Honey." *Food Additives & Contaminants: Part A* 34, no. 11 (2017): 1982–1989. https://doi.org/10.1080/19440049.2017.1347281.

Munoz-Pineiro, Maria Amalia, and Brigitte Toussaint. "'Fake Rice' on African and Asian Markets: Rumour or Evidence? Factsheet—December 2017." European Commission, 2018, JCR110625.

National Health Service. "The Research on Superfoods Is Exaggerated by the Media." In *Superfoods*, edited by Roman Espejo. Farmington Hills, MI: Greenhaven Press, 2016.

Nettle, Daniel, Clare Andrews, and Melissa Bateson. "Food Insecurity as a Driver of Obesity in Humans: The Insurance Hypothesis." *Behavioral and Brain Sciences* 40 (2017): e105. https://doi.org/10.1017/S0140525X16000947.

Newman, Lenore. *Lost Feast: Culinary Extinction and the Future of Food.* Toronto: ECW Press, 2019.

Niaz, Kamal, Jonathan Spoor, and Elizabeta Zaplatic. "Highlight Report: *Diploptera functata* (Cockroach) Milk as Next Superfood." *EXCLI Journal* 17 (2018): 721–723. https://doi.org/10.17179/excli2018-1437.

Nogués-Bravo, David, Jesús Rodríguez, Joaquín Hortal, Persaram Batra, and Miguel B. Araújo. "Climate Change, Humans, and the Extinction of the Woolly Mammoth." *PLoS Biology* 6, no. 4 (2008): 685–692. https://doi.org/10.1371/journal.pbio.0060079.

Nowak, Peter. *Sex, Bombs, and Burgers: How War, Pornography, and Fast Food Have Shaped Modern Technology.* Guilford, CT: Lyons Press, 2010.

Nukmal, Nismah, Suratman Umar, Sheila Puspita Amanda, and Mohammad Kanedi. "Effect of Styrofoam Waste Feeds on the Growth, Development and Fecundity of Mealworms (*Tenebrio molitor*)." *OnLine Journal of Biological Sciences* 18, no. 1 (2018): 24–28. https://doi.org/10.3844/ojbsci.2018.24.28.

Nuñez, Martin A., Sara Kuebbing, Romina D. Dimarco, and Daniel Simberloff. "Invasive Species: To Eat or Not to Eat, That Is the Question." *Conservation Letters* 5, no. 5 (2012): 334–341. https://doi.org/10.1111/j.1755-263x.2012.00250.x.

Ong Xiang An, Jason. "Future Typologies Part 1 | Restaurants | Now Serving: Future Foods." 2019. https://doi.org/10.13140/RG.2.2.22922.21448.

Oonincx, Dennis G. A. B., and Imke J. M. de Boer. "Environmental Impact of the Production of Mealworms as a Protein Source for Humans—A Life Cycle Assessment." *PLoS ONE* 7, no. 12 (2012). https://doi.org/10.1371/journal.pone.0051145.

Organisation for Economic Co-Operation and Development (OECD). *The Heavy Burden of Obesity: The Economics of Prevention.* Paris: OECD, 2019.

Orr, John Boyd, and David Lubbock. *The White Man's Dilemma.* 2nd ed. London: George Allen & Unwin, 1965. First published in 1953.

Östlund, Lars, Lisa Ahlberg, Olle Zackrisson, Ingela Bergman, and Steve Arno. "Bark-Peeling, Food Stress and Tree Spirits: The Use of Pine Inner Bark for Food in Scandinavia and North America." *Journal of Ethnobiology* 29, no. 1 (2009): 94–112. https://doi.org/10.2993/0278-0771-29.1.94.

Owen, Derek H., and David F. Katz. "A Review of the Physical and Chemical Properties of Human Semen and the Formulation of a Semen Simulant." *Journal of Andrology*, 2005, 26, no. 4 (2005): 459–469. https://doi.org/10.2164/jandrol.04104.

Pan, An, Vasanti Malik, and Frank B. Hu. "Exporting Diabetes to Asia: The Impact of Western-Style Fast Food." *Circulation*, 2, 10, no. 126 (2012): 163–165. https://doi.org/10.1161/CIRCULATIONAHA.112.115923.

Pfenniger, Alois, Rolf Vogel, Volker M. Koch, and Magnus Jonsson. "Performance Analysis of a Miniature Turbine Generator for Intracorporeal Energy Harvesting." *Artificial Organs* 38, no. 5 (2014): 68–81. https://doi.org/10.1111/aor.12279.

Phillips, Tony. "The Mysterious Smell of Moondust." NASA Science, January 30, 2006. https://science.nasa.gov/science-news/science-at-nasa/2006/30jan_smellofmoondust.

Photenhauer, Paul "Fotie." *Natural Harvest: A Collection of Semen-Based Recipes*. North Charleston, SC: Createspace Independent Publishing Platform, 2008.

Pollan, Michael. *The Omnivore's Dilemma: A Natural History of Four Meals*. New York: Penguin Books, 2006.

Rabbani, Ataul Goni, Shakila Faruque, Rakibul Hassan, and Nathu Ram Sarker. "Investigating the Existence of Artificial Eggs in Bangladesh and the Fact." *Journal of Applied Sciences* 19, no. 7 (2019): 701–707. https://doi.org/10.3923/jas.2019.701.707.

Ramachandraiah, Karna. "Potential Development of Sustainable 3D-Printed Meat Analogues: A Review." *Sustainability* 13, no. 2 (2021): 938. https://doi.org/10.3390/su13020938.

Ramos-Elorduy, Julieta. *Creepy Crawly Cuisine: The Gourmet Guide to Edible Insects*. Translated by Nancy Esteban. Rochester, VT: Park Street Press, 1998.

Reeves-Evison, Theo. *Ethics of Contemporary Art: In the Shadow of Transgression*. New York: Bloomsbury Visual Arts, 2020.

Relman, David A. "New Technologies, Human-Microbe Interactions, and the Search for Previously Unrecognized Pathogens." *Journal of Infectious Diseases* 186, no. 2 (2002): 254–258. https://doi.org/10.1086/344935.

Rettore, Anna, Róisín M. Burke, and Catherine Barry-Ryan. "Insects: A Protein Revolution for the Western Human Diet." In *Dublin Gastronomy Symposium, May 31–June 1, 2016*, 2016.

Ricciardi, Anthony. "Invasive Species." In *Encyclopedia of Sustainability Science and Technology*, edited by Robert A. Meyers, 161–178. New York: Springer, 2012.

Richardson, Matthew L., Robert F. Mitchell, Peter F. Reagel, and Lawrence Hanks. "Causes and Consequences of Cannibalism in Noncarnivorous Insects." *Annual Review of Entomology* 55, no. 1 (2010): 39–53.

Richey, Alexandra S., Brian F. Thomas, Min-Hui Lo, John T. Reager, James S. Famiglietti, Katalyn Voss, Sean Swenson, and Matthew Rodell. "Quantifying Renewable Groundwater Stress with Grace." *Water Resources Research* 51, no. 7 (2015): 5217–5238. https://doi.org/10.1002/2015wr017349.

Roche, Julian. "Farm Management." In *Agribusiness: An International Perspective*, 124–169. Abingdon, UK: Routledge, 2020.

Romero, Mariana Meneses. "Eating Human Cheese: The Lady Cheese Shop (Est. 2011)." *Feast Journal*, no. 3 (2016).

Rose, C., A. Parker, B. Jefferson, and E. Cartmell. "The Characterization of Feces and Urine: A Review of the Literature to Inform Advanced Treatment Technology." *Critical Reviews in Environmental Science and Technology* 45, no. 17 (2015): 1827–1879. https://doi.org/10.1080/10643389.2014.1000761.

Ross, Russell. "The Smooth Muscle Cell." *Journal of Cell Biology* 50, no. 1 (1971): 172–186. https://doi.org/10.1083/jcb.50.1.172.

Rudolf, Volker H. W., and Janis Antonovics. "Disease Transmission by Cannibalism: Rare Event or Common Occurrence?" *Proceedings of the Royal Society B: Biological Sciences* 274, no. 1614 (2007): 1205–1210. https://doi.org/10.1098/rspb.2006.0449.

Rumpho, Mary E., Elizabeth J. Summer, and James R. Manhart. "Solar-Powered Sea Slugs: Mollusc/Algal Chloroplast Symbiosis." *Plant Physiology* 123, no. 1 (2000): 29–38. https://doi.org/10.1104/pp.123.1.29.

Sánchez-Bayo, Francisco, and Kris A. G. Wyckhuys. "Worldwide Decline of the Entomofauna: A Review of Its Drivers." *Biological Conservation* 232 (2019): 8–27. https://doi.org/10.1016/j.biocon.2019.01.020.

Sandelin, Erik, and Unsworn Industries. "Eating E.T." In *Exploring the Animal Turn: Human-Animal Relations in Science, Society and Culture*, edited by Erika Andersson Cederholm, Björck Amelie, Kristina Jennbert, and Lönngren Ann-Sofie, 47–56. Lund, Sweden: Pufendorfinstitutet, 2014.

Savica, Vincenzo, Lorenzo A. Calò, Domenico Santoro, Paolo Monardo, Agostino Mallamace, and Guido Bellinghieri. "Urine Therapy through the Centuries." *Journal of Nephrology* 24, no. Suppl. 17 (2011): 123–125. https://doi.org/10.5301/JN.2011.6463.

Savoca, Matthew S., Chris W. Tyson, Michael McGill, and Christina J. Slager. "Odours from Marine Plastic Debris Induce Food Search Behaviours in a Forage Fish." *Proceedings of the Royal Society B: Biological Sciences* 284, no. 1860 (2017): 20171000. https://doi.org/10.1098/rspb.2017.1000.

Schedel, Hartmann, and Anton Koberger. *The Nuremberg Chronicle: A Facsimile of Hartmann Schedel's* Buch Der Chroniken, *Printed by Anton Koberger in 1493*. New York: Landmark Press, 1979.

Science Communication Unit. *Science for Environment Policy In-Depth Report: Sustainable Food*. Report produced for the European Commission DG Environment. Bristol: University of the West of England, 2013. http://ec.europa.eu/science-environment-policy.

Scrinis, Gyorgy. *Nutritionism: The Science and Politics of Dietary Advice*. New York: Columbia University Press, 2013.

Ségurel, Laure, and Céline Bon. "On the Evolution of Lactase Persistence in Humans." *Annual Review of Genomics and Human Genetics* 18, no. 1 (2017): 297–319. https://doi.org/10.1146/annurev-genom-091416-035340.

Sender, Ron, Shai Fuchs, and Ron Milo. "Revised Estimates for the Number of Human and Bacteria Cells in the Body." *PLoS Biology* 14, no. 8 (2016). https://doi.org/10.1371/journal.pbio.1002533.

Shen, Anqi. "'Being Affluent, One Drinks Wine': Wine Counterfeiting in Mainland China." *International Journal for Crime, Justice and Social Democracy* 7, no. 4 (2018): 16–32. https://doi.org/10.5204/ijcjsd.v7i4.1086.

Showalter, Elaine. *Hystories: Hysterical Epidemics and Modern Culture*. London: Picador, 1997.

Shrader, H. L. "The Chicken-of-Tomorrow Program: Its Influence on 'Meat-Type' Poultry Production." *Poultry Science* 31, no. 1 (1952): 3–10. https://doi.org/10.3382/ps.0310003.

Shridhar, G., N. Rajendra, H. Murigendra, P. Shridev, M. Prasad, M. A. Prasad, S. Prasad, et al. "Modern Diet and Its Impact on Human Health." *Journal of Nutrition & Food Sciences* 5, no. 6 (2015): 430. https://doi.org/10.4172/2155-9600.1000430.

Sillman, Jani, Lauri Nygren, Helena Kahiluoto, Vesa Ruuskanen, Anu Tamminen, Cyril Bajamundi, Marja Nappa, et al. "Bacterial Protein for Food and Feed Generated via Renewable Energy and Direct Air Capture of CO_2: Can It Reduce Land and Water Use?" *Global Food Security* 22 (2019): 25–32. https://doi.org/10.1016/j.gfs.2019.09.007.

Smith, Andrew F. *Encyclopedia of Junk Food and Fast Food.* Westport, CT: Greenwood Press, 2006.

Smith, Andrew F. *Potato: A Global History.* London: Reaktion Books, 2011.

Smith, Madeleine, David C. Love, Chelsea M. Rochman, and Roni A. Neff. "Microplastics in Seafood and the Implications for Human Health." *Current Environmental Health Reports* 5, no. 3 (2018): 375–386. https://doi.org/10.1007/s40572-018-0206-z.

Soares, Susana, and Andrew Forkes. "Insects Au Gratin: An Investigation into the Experiences of Developing a 3D Printer That Uses Insect Protein Based Flour as a Building Medium for the Production of Sustainable Food." In *DS 78: Proceedings of the 16th International Conference on Engineering and Product Design Education (E&PDE14), Design Education and Human Technology Relations, University of Twente, the Netherlands, September 4–5, 2014,* edited by Erik Bohemia, Arthur Eger, Wouter Eggink, Ahmed Kovacevic, Brian Parkinson, and Wessel Wits, 426–331. Design Society, 2014.

Specht, Kathrin, Felix Zoll, Henrike Schümann, Julia Bela, Julia Kachel, and Marcel Robischon. "How Will We Eat and Produce in the Cities of the Future? From Edible Insects to Vertical Farming—A Study on the Perception and Acceptability of New Approaches." *Sustainability* 11, no. 4315 (2019). https://doi.org/10.3390/su11164315.

Stead, Selina M., and Lindsey Laird, eds. *Handbook of Salmon Farming.* London: Springer, 2002.

Steffen, Will, Paul J. Crutzen, and John R. McNeill. "The Anthropocene: Are Humans Now Overwhelming the Great Forces of Nature?" *AMBIO: A Journal of the Human Environment* 36, no. 8 (2007): 614–621. https://doi.org/btvrzb.

Steinberg, Lisa M., Rachel E. Kronyak, and Christopher H. House. "Coupling of Anaerobic Waste Treatment to Produce Protein- and Lipid-Rich Bacterial Biomass." *Life Sciences in Space Research* 15 (2017): 32–42. https://doi.org/10.1016/j.lssr.2017.07.006.

Stenslie, Ståle. *Virtual Touch: A Study of the Use and Experience of Touch in Artistic, Multimodal and Computer-Based Environments.* Oslo: Oslo School of Architecture and Design, 2010.

Strahler, Jana, Andrea Hermann, Bertram Walter, and Rudolf Stark. "Orthorexia Nervosa: A Behavioral Complex or a Psychological Condition?" *Journal of Behavioral Addictions* 7, no. 4 (2018): 1143–1156. https://doi.org/10.1556/2006.7.2018.129.

Sun, Zhejun, Hao Jiang. "Nutritive Evaluation of Earthworms as Human Food." In *Future Foods*, edited by Heimo Juhani Mikkola, 127–141. Rijeka, Croatia: IntechOpen, 2017.

Symons, Michael. *A History of Cooks and Cooking*. Urbana: University of Illinois Press, 2003. First published in 1945.

Taleb, Nassim Nicholas. *The Black Swan: The Impact of the Highly Improbable.* New York: Random House, 2007.

Tandon, Shalini, Niranjan Kolekar, and Rakesh Kumar. "Water and Energy Footprint Assessment of Bottled Water Industries in India." *Natural Resources* 5 (2014): 68–72. https://doi.org/10.4236/nr.2014.52007.

Taylor, Geoffrey, and Nanxi Wang. "Entomophagy and Allergies: A Study of the Prevalence of Entomophagy and Related Allergies in a Population Living in North-Eastern Thailand." *Bioscience Horizons: The International Journal of Student Research* 11 (2018). https://doi.org/10.1093/biohorizons/hzy003.

Tharp, Bruce M., and Stephanie M. Tharp. *Discursive Design: Critical, Speculative, and Alternative Things.* Cambridge, MA: MIT Press, 2019.

Thomas, Chris D. "The Sixth Mass Genesis? New Species Are Coming into Existence Faster than Ever Thanks to Humans." The Conversation, October 6, 2019. https://theconversation.com/the-sixth-mass-genesis-new-species-are-coming-into-existence-faster-than-ever-thanks-to-humans-80527.

Tieken, S. M., H. J. Leidy, A. J. Stull, R. D. Mattes, R. A. Schuster, and W. W. Campbell. "Effects of Solid versus Liquid Meal-Replacement Products of Similar Energy Content on Hunger, Satiety, and Appetite-Regulating Hormones in Older Adults." *Hormone and Metabolic Research* 39, no. 5 (2007): 389–394. https://doi.org/10.1055/s-2007-976545.

Tilman, David, Christian Balzer, Jason Hill, and Belinda L. Befort. "Global Food Demand and the Sustainable Intensification of Agriculture." *PNAS* 108, no. 50 (2011): 20260–20264. https://doi.org/10.1073/pnas.1116437108.

Todorović, Zoran. "Human Gourmet." Interview by Zane Cerpina. *EE Experimental Emerging Art* no. 3 (2018): 50–53.

Tunick, Michael H. *The Science of Cheese.* New York: Oxford University Press, 2014.

Tunio, Mazhar H., Jianmin Gao, Sher A. Shaikh, Imran A. Lakhiar, Waqar A. Qureshi, Kashif A. Solangi, and Farman A. Chandio. "Potato Production in Aeroponics: An Emerging Food Growing System in Sustainable Agriculture for Food Security." *Chilean Journal of Agricultural Research* 80, no. 1 (2020): 118–132. https://doi.org/10.4067/s0718-58392020000100118.

Turvey, S. T., and C. L. Risley. "Modelling the Extinction of Steller's Sea Cow." *Biology Letters* 2, no. 1 (2006): 94–97. https://doi.org/10.1098/rsbl.2005.0415.

United Nations Department of Economic and Social Affairs, Population Division. *World Urbanization Prospects: The 2018 Revision (ST/ESA/SER.A/420).* New York: United Nations, 2019.

United Nations Educational, Scientific, and Cultural Organization (UNESCO). "Futures Literacy: An Essential Competency for the 21st Century." Accessed July 29, 2021. https://en.unesco.org/futuresliteracy.

United States Department of the Army. *US Army Field Manual 3-05.70: Survival.* Washington, DC: Government Printing Office, 2002.

University of the Arts London. *Material Futures.* London: University of the Arts London, 2016. https://issuu.com/csmtime/docs/mamf_course_catalogue_2016/23.

Valmalette, Jean Christophe, Aviv Dombrovsky, Pierre Brat, Christian Mertz, Maria Capovilla, and Alain Robichon. "Light-Induced Electron Transfer and ATP Synthesis in a Carotene Synthesizing Insect." *Scientific Reports* 2, no. 1 (2012). https://doi.org/10.1038/srep00579.

Van Den Bos Verma, Monika, Linda de Vreede, Thom Achterbosch, and Martine M. Rutten. "Consumers Discard a Lot More Food than Widely Believed: Estimates of Global Food Waste Using an Energy Gap Approach and Affluence Elasticity of Food Waste." *PLoS ONE* 15, no. 2 (2020): 1–4. https://doi.org/10.1371/journal.pone.0228369.

Van Huis, Arnold, and Dennis G. Oonincx. "The Environmental Sustainability of Insects as Food and Feed: A Review." *Agronomy for Sustainable Development* 37, no. 5 (2017): 43. https://doi.org/10.1007/s13593-017-0452-8.

Van Huis, Arnold, Joost Van Itterbeeck, Harmke Klunder, Esther Mertens, Afton Halloran, Giulia Muir, and Paul Vantomme. *Edible Insects: Future Prospects for Food and Feed Security.* Rome: Food and Agriculture Organization of the United Nations, 2013.

Van Huis, Arnold. "Did Early Humans Consume Insects?" *Journal of Insects as Food and Feed* 3, no. 3 (2017): 161–163. https://doi.org/10.3920/jiff2017.x006.

Van Huis, Arnold. "Edible Insects Contributing to Food Security?" *Agriculture & Food Security* 4, no. 1 (2015): 20. https://doi.org/10.1186/s40066-015-0041-5.

Van Huis, Arnold. "Potential of Insects as Food and Feed in Assuring Food Security." *Annual Review of Entomology* 58, no. 1 (2013): 563–583. https://doi.org/10.1146/annurev-ento-120811-153704.

Van Mensvoort, Koert, and Hendrik-Jan Grievink. *The In Vitro Meat Cookbook.* Amsterdam: BIS and Next Nature Network, 2014.

Van Mensvoort, Koert, and Hendrik-Jan Grievink, eds. *Next Nature: Nature Changes Along with Us.* Amsterdam: Next Nature Network, 2015.

Varelas, Vassileios. "Food Wastes as a Potential New Source for Edible Insect Mass Production for Food and Feed: A Review." *Fermentation* 5, no. 3 (2019): 81. https://doi.org/10.3390/fermentation5030081.

Virilio, Paul. "Surfing the Accident." Interview by Andreas Ruby. In *The Art of the Accident: Merging of Art, Architecture and Media Technology*, edited by Bart Lootsma, Joke Brouwer and Arjen Mulder (Rotterdam: NAI and V2_Organization, 1998), 30–47.

Vitkalova, Anna V., Limin Feng, Alexander N. Rybin, Brian D. Gerber, Dale G. Miquelle, Tianming Wang, Haitao Yang, Elena I. Shevtsova, Vladimir V. Aramilev, and Jianping Ge. "Transboundary Cooperation Improves Endangered Species Monitoring and Conservation Actions: A Case Study of the Global Population of Amur Leopards." *Conservation Letters* 11, no. 5 (2018): e12574. https://doi.org/10.1111/conl.12574.

Waffle, Alexander D., Robert C. Corry, Terry J. Gillespie, and Robert D. Brown. "Urban Heat Islands as Agricultural Opportunities: An Innovative Approach." *Landscape and Urban Planning* 161 (2017): 103–114. https://doi.org/10.1016/j.landurbplan.2017.01.010.

Wang, Liming, and John Davis. *China's Grain Economy: The Challenge of Feeding More than a Billion.* London: Routledge, 2018.

Waters, C. N., J. Zalasiewicz, C. Summerhayes, A. D. Barnosky, C. Poirier, A. Ga Uszka, A. Cearreta, et al. "The Anthropocene Is Functionally and Stratigraphically Distinct from the Holocene." *Science* 351, no. 6269 (2016): 137–147. https://doi.org/10.1126/science.aad2622.

Wilcox, Howard A. *Hothouse Earth.* New York: Praeger, 1975.

Williams, H. K. "Religious Education." *The Biblical World* 53, no. 1 (1919): 78–81.

Williford, Anna, Barbara Stay, and Debashish Bhattacharya. "Evolution of a Novel Function: Nutritive Milk in the Viviparous Cockroach, *Diploptera punctata*." *Evolution and Development* 6, no. 2 (2004): 67–77. https://doi.org/10.1111/j.1525-142x.2004.04012.x.

Wiss, David A., Nicole Avena, and Pedro Rada. "Sugar Addiction: From Evolution to Revolution." *Frontiers in Psychiatry* 9, no. 545 (2018): 545. https://doi.org/10.3389/fpsyt.2018.00545.

Wood, Gillen D'Arcy. *Tambora: The Eruption That Changed the World*. Princeton, NJ: Princeton University Press, 2014.

World Bank Group. *An Overview of Links between Obesity and Food Systems*. Washington, DC: World Bank, 2017.

World Economic Forum. "The New Plastics Economy: Rethinking the Future of Plastics." World Economic Forum, 2016.

Wright, Ronald. "Ronald Wright: Can We Still Dodge the Progress Trap?" *The Tyee*, September 20, 2019. https://thetyee.ca/Analysis/2019/09/20/Ronald-Wright-Can-We-Dodge-Progress-Trap/.

Wright, Ronald. *A Short History of Progress*. Toronto: House of Anansi Press, 2011. First published in 2004.

Yang, Bin, Jianwu Wang, Bo Tang, Yufang Liu, Chengdong Guo, Penghua Yang, Tian Yu, et al. "Characterization of Bioactive Recombinant Human Lysozyme Expressed in Milk of Cloned Transgenic Cattle." *PLoS ONE* 6, no. 3 (2011). https://doi.org/10.1371/journal.pone.0017593.

Yang, Jun, Yu Yang, Wei-Min Wu, Jiao Zhao, and Lei Jiang. "Evidence of Polyethylene Biodegradation by Bacterial Strains from the Guts of Plastic-Eating Waxworms." *Environmental Science & Technology* 48, no. 23 (2014): 13776–13784. https://doi.org/10.1021/es504038a.

Yates, Joe, Megan Deeney, Howard White, Edward Joy, Sofia Kalamatianou, and Suneetha Kadiyala. "PROTOCOL: Plastics in the Food System: Human Health, Economic and Environmental Impacts. A Scoping Review." *Campbell Systematic Reviews* 15, no. 1–2 (2019): e1033. https://doi.org/10.1002/cl2.1033.

Yoshida, Shosuke, Kazumi Hiraga, Toshihiko Takehana, Ikuo Taniguchi, Hironao Yamaji, Yasuhito Maeda, Kiyotsuna Toyohara, Kenji Miyamoto, Yoshiharu Kimura, and Kohei Oda. "A Bacterium That Degrades and Assimilates Poly(Ethylene Terephthalate)." *Science* 351, no. 6278 (2016): 1196–1199. https://doi.org/10.1126/science.aad6359.

Young, Lisa R., and Marion Nestle. "Portion Sizes and Obesity: Responses of Fast-Food Companies." *Journal of Public Health Policy* 28, no. 2 (2007): 238–248. https://doi.org/10.1057/palgrave.jphp.3200127.

Zakaria-Runkat, Fransiska, Wanchai Worawattanamateekul, and Ong-Ard Lawhavinit. "Production of Fish Serum Products as Substitute for Fetal Bovine Serum in Hybridoma Cell Cultures from Surimi Industrial Waste." *Kasetsart Journal: Natural Science* 40 (2006): 198–205.

Zalasiewicz, Jan, Colin N. Waters, Juliana A. Ivar do Sul, Patricia L. Corcoran, Anthony D. Barnosky, Alejandro Cearreta, Matt Edgeworth, et al. "The Geological Cycle of Plastics and Their Use as a Stratigraphic Indicator of the Anthropocene." *Anthropocene* 13 (2016): 4–17. https://doi.org/10.1016/j.ancene.2016.01.002.

Zalasiewicz, Jan, Colin N. Waters, Mark Williams, Anthony D. Barnosky, Alejandro Cearreta, Paul Crutzen, Erle Ellis, et al. "When Did the Anthropocene Begin? A Mid-Twentieth Century Boundary Level Is Stratigraphically Optimal." *Quaternary International* 383 (2015): 196–203. https://doi.org/10.1016/j.quaint.2014.11.045.

Zeldovich, Lina. "A History of Human Waste as Fertilizer." JSTOR Daily, November 18, 2019. https://daily.jstor.org/a-history-of-human-waste-as-fertilizer/.

Zhu, Gengping, Javier Gutierrez Illan, Chris Looney, and David W. Crowder. "Assessing the Ecological Niche and Invasion Potential of the Asian Giant Hornet." *PNAS* 117, no. 40 (2020): 24646–24648. https://doi.org/10.1101/2020.05.25.115311.

Žižek, Slavoj. "Slavoj Žižek: 'Humanity Is OK, but 99% of People Are Boring Idiots.'" Interview by Decca Aitkenhead. *The Guardian*, June 10, 2012. https://www.theguardian.com/culture/2012/jun/10/slavoj-zizek-humanity-ok-people-boring.

Zucoloto, Fernando Sérgio. "Evolution of the Human Feeding Behavior." *Psychology & Neuroscience* 4, no. 1 (2011): 131–141. http://dx.doi.org/10.3922/j.psns.2011.1.015.

Zuolo, Federico, Chiara Testino, and Emanuela Ceva. "The Challenges of Dietary Pluralism." In *The Routledge Handbook of Food Ethics*, edited by Mary Rawlinson and Caleb Ward, 93–102. Abington, UK: Routledge, 2017.

Illustration Credits

Figure 2.1. © Tattfoo Tan, *New Earth Meal Ready to Eat (NEMRE)*, 2013. Photograph courtesy of Tattfoo Tan.

Figure 2.2. © Jimmy Tang, *Future Food Hack*, 2015. Photograph courtesy of Jimmy Tang.

Figure 2.3. © Jimmy Tang, *Future Food Hack*, 2015. Photograph courtesy of Jimmy Tang.

Figure 2.4. © Paul Gong, *Human Hyena*, 2014. Photograph courtesy of Paul Gong. Photograph by Andrew Kan.

Figure 2.5. © Paul Gong, *Human Hyena*, 2014. Photograph courtesy of Paul Gong. Photograph by Andrew Kan.

Figure 2.6. © Gints Gabrans, *FOOOD*, 2014. Photograph courtesy of Gints Gabrans.

Figure 2.7. © Gints Gabrans, *FOOOD*, 2014. Photograph courtesy of Gints Gabrans.

Figure 3.1. © Nonhuman Nonsense, *Pink Chicken Project*, 2017. Photograph courtesy of Nonhuman Nonsense.

Figure 3.2. © Nonhuman Nonsense, *Pink Chicken Project*, 2017. Photograph courtesy of Nonhuman Nonsense.

Figure 3.3. © Miriam Simun and Miriam Songster, *GhostFood*, 2013. Photograph courtesy of Miriam Simun and Miriam Songster. Photograph by Miriam Simun.

Figure 3.4. © Miriam Simun and Miriam Songster, *GhostFood*, 2013. Photograph courtesy of Miriam Simun and Miriam Songster. Photograph by Miriam Simun.

Figure 3.5. © The Center for Genomic Gastronomy, *Cobalt 60 Sauce*, 2013. Photograph courtesy of The Center for Genomic Gastronomy.

Figure 3.6. © The Center for Genomic Gastronomy, *Cobalt 60 Sauce*, 2013. Photograph courtesy of The Center for Genomic Gastronomy.

Figure 3.7. © LIVIN Studio, *Fungi Mutarium*, 2014. Photograph courtesy of © LIVIN Studio. www.livinstudio.com.

Figure 3.8. © LIVIN Studio, *Fungi Mutarium*, 2014. Photograph courtesy of © LIVIN Studio. www.livinstudio.com.

Figure 3.9. © The Center for Genomic Gastronomy, *Smog Tasting*, 2011. Photograph courtesy of The Center for Genomic Gastronomy.

Figure 3.10. Jon Cohrs and Morgan Levy, *Alviso's Medicinal All-Salt*, 2010. Photograph courtesy of Jon Cohrs.

Figure 4.1. © Oron Catts and Ionat Zurr, *The Semi-living Steak*, 2000, part of the Tissue, Culture & Art (TC&A) Project, *Tissue Engineered Steak No. 1: A Study for Disembodied Cuisine*. Prenatal sheep skeletal muscle and degradable PGA polymer scaffold. Photograph courtesy of the Tissue, Culture & Art Project (Oron Catts and Ionat Zurr). TC&A is hosted at SymbioticA, School of Anatomy, Physiology and Human Biology, University of Western Australia. Photograph courtesy of the Tissue, Culture & Art Project (Oron Catts and Ionat Zurr).

Figure 4.2. © Oron Catts and Ionat Zurr, *The Semi-living Steak*, 2000, part of the Tissue, Culture & Art (TC&A) Project, *Tissue Engineered Steak No.1: A Study for Disembodied Cuisine*. Prenatal sheep skeletal muscle and degradable PGA polymer scaffold. Photograph courtesy of the Tissue, Culture & Art Project (Oron Catts and Ionat Zurr). TC&A is hosted at SymbioticA, School of Anatomy, Physiology and Human Biology, University of Western Australia. Photograph courtesy of the Tissue, Culture & Art Project (Oron Catts and Ionat Zurr).

Figure 4.3. © Kuang-Yi Ku, *Tiger Penis Project*, 2018. Photo courtesy of Kuang-Yi Ku. Photograph © Ronald Smits / Design Academy Eindhoven.

Figure 4.4. © Oron Catts, Ionat Zurr, and Robert Foster, *Stir Fly: Nutrient Bug 1.0*, 2016, part of the Tissue, Culture & Art (TC&A)Project. Photograph courtesy of the Tissue, Culture & Art Project (Oron Catts and Ionat Zurr). *TC&A* is hosted at SymbioticA, School of Anatomy, Physiology and Human Biology, University of Western Australia. Photograph courtesy of the Tissue, Culture & Art Project (Oron Catts and Ionat Zurr).

Figure 4.5. *In Vitro Oysters* from *The In Vitro Meat Cookbook* (2014) by Koert van Mensvoort and Hendrik-Jan Grievink. © Next Nature Network and BIS Publishers. Photograph courtesy of Next Nature Network.

Figure 4.6. *Knitted Meat* by Alberto Gruarin from *The In Vitro Meat Cookbook* (2014) by Koert van Mensvoort and Hendrik-Jan Grievink. © Next Nature Network and BIS Publishers. Photograph courtesy of Next Nature Network.

Figure 4.7. © dentsu / Open Meals, *Digital Oden*, 2018. Photograph courtesy of dentsu / Open Meals.

Figure 4.8. © dentsu / Open Meals, *Digital Oden*, 2018. Photograph courtesy of dentsu / Open Meals.

Figure 5.1. © Christina Agapakis and Sissel Tolaas, *Selfmade*, 2013. Photograph courtesy of Christina Agapakis and Sissel Tolaas.

Figure 5.2. © Christina Agapakis and Sissel Tolaas, *Selfmade*, 2013. Photograph courtesy of Christina Agapakis and Sissel Tolaas.

Figure 5.3. © Miriam Simun, *Lady Cheese Shop*, 2011. Making human cheese. Video frame.

Figure 5.4. © Miriam Simun, *Lady Cheese Shop*, 2011. Photo by Shimpel Takeda.

Figure 5.5. © James Gilpin, *Gilpin Family Whisky*, 2010. Photograph courtesy of James Gilpin.

Figure 5.6. © Marko Marković, *Selfeater/Hunger*, from the performance series *Selfeater*, 2009. Photograph courtesy of Marko Marković. Photograph by Silvijo Selman.

Figure 5.7. © Theresa Schubert, *mEat me*, 2020. Photograph courtesy of Theresa Schubert. Photograph by Tina Lagler / Kapelica Gallery Archive.

Figure 5.8. © Theresa Schubert, *mEat me*, 2020. Photograph courtesy of Theresa Schubert. Photograph by Hana Jošić / Kapelica Gallery Archive.

Figure 5.9. © The Center for Genomic Gastronomy, *To Flavour Our Tears*, 2016. Photograph courtesy of The Center for Genomic Gastronomy.

Figure 6.1. The Anthropocene Kitchen, *Ant(i) Pasti*, 2017. Photo courtesy of Zane Cerpina and Stahl Stenslie. Photograph by Janis Viksna.

Figure 6.2. © *SexyFood*, 2014–2016. Packaging design by Atelier Design (Steven van Boxtel). Photograph courtesy of SexyFood.

Figure 6.3. Original idea by Susana Soares, project coordination by Susana Soares and Andrew Forkes, *Insects Au Gratin*, 2011. Photograph courtesy of Susana Soares.

Figure 6.4. © Nordic Food Lab and the Cambridge Distillery, *Anty Gin*, 2013. Photograph courtesy of the Cambridge Distillery.

Figure 6.5. © LIVIN Farms, *Hive*, 2018. Photograph courtesy © LIVIN Farms. www.livinfarms.com.

Figure 6.6. © Marc Paulusma, *BoeteBurger Project*, 2019. Photograph courtesy of Marc Paulusma.

Figure 6.7. © Marc Paulusma, *BoeteBurger Project,* 2019. Photograph courtesy of Marc Paulusma.

Figure 6.8. © Wataru Kobayashi, *BUGBUG*, 2016. Textured spoons designed for grinding insects. Photograph courtesy of Wataru Kobayashi.

Figure 6.9. © Wataru Kobayashi, *BUGBUG*, 2016. A cutlery set for eating insects. Photograph courtesy of Wataru Kobayashi.

Figure 7.1. Anthropocene Kitchen, *Meal-in-a-Pill*, 2017. Photograph courtesy of Anthropocene Kitchen.

Figure 7.2. © Andreas Greiner, *Monument for the 308: Monument for a Contemporary Dinosaur (Ross 308)*, 2016. Installation view at Berlinische Galerie. Photograph by Theo Bitzer, 2016.

Figure 7.3. © Paul Gong, *The Cow of Tomorrow*, 2015. Photograph by Lydia Chang.

Figure 7.4. Michael Burton and Michiko Nitta, *Near Future Algae Symbiosis Suit: Prototype* (2010). All content copyright reserved © 2010 Burton Nitta (Michael Burton and Michiko Nitta). Photograph courtesy of Michael Burton and Michiko Nitta.

Figure 7.5. © dentsu / Open Meals, *Sushi Singularity*, 2019. Photograph courtesy of dentsu / Open Meals.

Figure 7.6. © dentsu / Open Meals, *Sushi Singularity*, 2019. Photograph courtesy of dentsu / Open Meals.

Figure 8.1. © Helmut Smits, *The Real Thing*, 2010–2017. Photograph courtesy of Lotte Stekelenburg.

Figure 8.2. © Helmut Smits, *The Real Thing*, 2010–2017. Photograph courtesy of Lotte Stekelenburg.

Figure 8.3. © Space10, *Tomorrow's Meatball: A Visual Exploration of Future Food*, 2015. Photograph by Lukas Renlund.

Figure 8.4. © Henry Hargreaves and Caitlin Levin, *Deep-Fried Gadgets*, 2013. Photograph courtesy of Henry Hargreaves and Caitlin Levin.

Figure 9.1. © The Center for Genomic Gastronomy, *The De-extinction Deli*, 2013. Photograph courtesy of The Center for Genomic Gastronomy.

Figure 9.2. © The Center for Genomic Gastronomy, *The De-extinction Deli*, 2013. Photograph courtesy of The Center for Genomic Gastronomy.

Figure 9.3. Anthropocene Kitchen. *Anthropocene Feast: Eating the Last Dragon*, 2017. Photograph courtesy of Anthropocene Kitchen. Photograph by Janis Viksna.

Figure 9.4. Anthropocene Kitchen. *Anthropocene Feast: Eating the Last Dragon*, 2017. Photograph courtesy of Anthropocene Kitchen. Photograph by Janis Viksna.

Figure 9.5. © Terje Östling and Unsworn Industries, *Eating E.T.—Mock Alien BBQ*, 2014. Photograph courtesy of Terje Östling and Unsworn Industries. Photograph by Bengt Pettersson.

Figure 9.6. © Terje Östling and Unsworn Industries, *Eating E.T.—Mock Alien BBQ*, 2014. Photograph courtesy of Terje Östling and Unsworn Industries. Photograph by Antti Ahonen.

Index

Note: Page numbers in italics refer to artwork.